FOREWORD

The closure of a deep geological repository for radioactive waste represents a transition from active management of the facility to passive safety. A convincing safety case addressing post-closure safety is required in order to support a positive decision to move on to the next step in the planning and development of such repositories. The safety case must provide evidence for safety at all relevant times subsequent to closure. Relevant times extend into the future for as long the radiotoxicity associated with the repository represents a hazard to human health and the environment. Depending on the half-lives and activities of the radionuclides under consideration, the safety case may thus need to address very long periods of time.

Post-closure safety is typically provided by a system of multiple barriers that generally includes an engineered barrier system and the geological setting (the natural barrier or barriers). The significance of particular barriers to system performance, the safety functions that the barriers provide, and uncertainties and perturbing phenomena that affect them, vary over the time period of concern in a safety case. One example of uncertainty is the relatively complex phenomena that may occur during an initial, transient phase subsequent to closure (e.g. resaturation phenomena and chemical reactions). Thus, some phenomena and uncertainties may be significant between certain time intervals and be of relatively little concern at other times.

The issue of how to handle phenomena and uncertainties that are characterised by widely different timescales is of concern to all national programmes. The most appropriate ways of quantifying performance or safety may also vary with time (the most common safety indicators are dose and risk, but these may be complemented by a number of other possible indicators). Thus, it may be convenient to divide the post-closure period into a number of intervals (time frames), that are characterised by particular types of phenomena or uncertainties, and for which particular types of performance or safety indicators are most suitable. This approach may also help with communicating and discussing the safety case with a wide range of audiences.

The various issues described above provided the motivation for the IGSC (Integration Group for the Safety Case of the OECD Nuclear Energy Agency) to support and organise a workshop entitled "Handling of timescales in assessing post-closure safety". It was held in Paris on 16-18 April 2002 and hosted by the French Institute for Radiological Protection and Nuclear Safety (IRSN). The main objective of the workshop was to identify and discuss approaches related to and work done on timescales issues within national waste management programmes, and in the context of assessing post-closure safety of deep geological repositories.

These proceedings include the presentations made at the workshop as well as a summary of the discussions held.

Acknowledgements

On behalf all the participants, the NEA wishes to express its gratitude to the French Institute for Radioprotection and Nuclear Safety (IRSN) for their hospitality and contribution to the organisation and success of the workshop.

The NEA is also very grateful to the members of the Programme Committee (PC) for their help in setting up, running the workshop and providing the proceedings:

- Peter De Preter, Chairman (ONDRAF/NIRAS, Belgium),
  Risto Paltemaa (STUK, Finland),
  Lise Griffault (ANDRA, France),
  Didier Gay (IRSN, France),
  Sylvie Voinis (NEA, France),
  Klaus-Jürgen Röhlig (GRS mbH, Germany),
  Hiroyuki Umeki (NUMO, Japan),
  Jürg Schneider (Nagra, Switzerland),
  Lucy Bailey and Anna Littleboy (Nirex Ltd, UK),
  Paul Smith (SAM Ltd, UK).

- The chairpersons and rapporteurs of the plenary and working group sessions who were very active in the animation and reporting of discussions (see PART C).

- The speakers and authors of papers for their contribution (see PART C).

- All the participants for their very active and constructive participation.

The analysis of the outcome of the workshop and technical summary has been prepared by Paul Smith (SAM Ltd, UK) based on discussions at the plenary session and the reports from the rapporteurs of each working group sessions and was reviewed by the Programme Committee. It also takes into account the comments made by chairpersons and rapporteurs.

The Handling of Timescales in Assessing Post-closure Safety of Deep Geological Repositories

**Workshop Proceedings
Paris, France
16-18 April 2002**

Hosted by
The French Institute for Radiological Protection
and Nuclear Safety (IRSN)

NUCLEAR ENERGY AGENCY
ORGANISATION FOR ECONOMIC CO-OPERATION AND DEVELOPMENT

ORGANISATION FOR ECONOMIC CO-OPERATION AND DEVELOPMENT

Pursuant to Article 1 of the Convention signed in Paris on 14th December 1960, and which came into force on 30th September 1961, the Organisation for Economic Co-operation and Development (OECD) shall promote policies designed:

- to achieve the highest sustainable economic growth and employment and a rising standard of living in Member countries, while maintaining financial stability, and thus to contribute to the development of the world economy;
- to contribute to sound economic expansion in Member as well as non-member countries in the process of economic development; and
- to contribute to the expansion of world trade on a multilateral, non-discriminatory basis in accordance with international obligations.

The original Member countries of the OECD are Austria, Belgium, Canada, Denmark, France, Germany, Greece, Iceland, Ireland, Italy, Luxembourg, the Netherlands, Norway, Portugal, Spain, Sweden, Switzerland, Turkey, the United Kingdom and the United States. The following countries became Members subsequently through accession at the dates indicated hereafter: Japan (28th April 1964), Finland (28th January 1969), Australia (7th June 1971), New Zealand (29th May 1973), Mexico (18th May 1994), the Czech Republic (21st December 1995), Hungary (7th May 1996), Poland (22nd November 1996), Korea (12th December 1996) and the Slovak Republic (14 December 2000). The Commission of the European Communities takes part in the work of the OECD (Article 13 of the OECD Convention).

NUCLEAR ENERGY AGENCY

The OECD Nuclear Energy Agency (NEA) was established on 1st February 1958 under the name of the OEEC European Nuclear Energy Agency. It received its present designation on 20th April 1972, when Japan became its first non-European full Member. NEA membership today consists of 28 OECD Member countries: Australia, Austria, Belgium, Canada, Czech Republic, Denmark, Finland, France, Germany, Greece, Hungary, Iceland, Ireland, Italy, Japan, Luxembourg, Mexico, the Netherlands, Norway, Portugal, Republic of Korea, Slovak Republic, Spain, Sweden, Switzerland, Turkey, the United Kingdom and the United States. The Commission of the European Communities also takes part in the work of the Agency.

The mission of the NEA is:

- to assist its Member countries in maintaining and further developing, through international co-operation, the scientific, technological and legal bases required for a safe, environmentally friendly and economical use of nuclear energy for peaceful purposes, as well as
- to provide authoritative assessments and to forge common understandings on key issues, as input to government decisions on nuclear energy policy and to broader OECD policy analyses in areas such as energy and sustainable development.

Specific areas of competence of the NEA include safety and regulation of nuclear activities, radioactive waste management, radiological protection, nuclear science, economic and technical analyses of the nuclear fuel cycle, nuclear law and liability, and public information. The NEA Data Bank provides nuclear data and computer program services for participating countries.

In these and related tasks, the NEA works in close collaboration with the International Atomic Energy Agency in Vienna, with which it has a Co-operation Agreement, as well as with other international organisations in the nuclear field.

TABLE OF CONTENTS

PART A

SYNTHESIS OF THE WORKSHOP

1.   Introduction

1.1   Background

The implementation of a deep geological repository programmes requires safety assessments to be carried out, in order to support a safety case, defined as [1]:

> "… a collection of arguments …
> in support of the long-term safety of the repository".

Deep geological repositories are sited and designed to protect humans and the environment from the hazards associated with radioactive waste. The strategy adopted is, for an initial period at least, one of "concentrate and contain". Radioactive material is isolated from human beings and the surface environment and radioactive decay greatly reduces the level of activity before any release of radioactivity to the surface environment can occur. A repository safety case needs to address the period for which it is possible to contain the radioactivity, as well as later times, when arguments are also sought that any releases that occur are sufficiently small, or sufficiently diluted and dispersed, that the concentrations of radioactivity to which they give rise pose no hazard. Generally speaking, repositories for spent fuel and high-level waste are sited and designed to provide a longer period of containment than repositories for low – and intermediate – level wastes. This is reasonable given the lower concentrations of radioactivity in repositories for low – and intermediate– level wastes.

Safety assessments, as currently practised, require an understanding of the evolution of the repository and its environment, often over timescales ranging from, say, a few tens or hundreds of years for transient processes associated with the engineered barrier system to possibly millions of years for changes within the geological environment. Furthermore, safety is usually assessed in terms of the primary indicators of dose and risk and, in order to evaluate these indicators, assumptions must be made regarding the habits of potentially exposed groups (e.g. diet, lifestyle and land use), and these may change over timescales of just a few years. The need to deal with such a wide range of timescales gives rise to a range of issues regarding the methods and presentation of safety assessments. The issue of why, and to what degree, it is necessary to assess safety over very long timescales also needs to be considered.

1.2   The aims and organisation of the workshop

Issues related to timescales are of interest in all countries considering the development of deep geological repositories. Moreover they are of interest to those charged with carrying out repository safety assessments, those responsible for reviewing such assessments and those responsible for setting standards against which the safety of a deep geological repository should be judged. In view of the common interest on the part of both regulators and implementers of repositories, the Nuclear Energy Agency of the OECD organised a workshop, hosted by the French Institute for Radiological Protection and Nuclear Safety (IRSN), to explore and discuss the relevant issues.

The aims of the workshop were:

- to identify where the technical community now stands with respect to these issues, both with respect to safety assessment approaches and regulatory views;

- to establish what are the common elements in approaches and views in different national programmes and what, if any, are the differences; and

- to suggest possible further work that might be useful to clarify issues and improve the approaches for dealing with them.

A plenary session held on the first day of the workshop, addressed the first of these aims, with presentations designed to illustrate current approaches and views. Four Working Groups held discussions throughout the next day and a half, covering between them the following detailed Technical Topics, and considering all three of the workshop aims.

Technical Topic A: The Different Timescales *versus* the Regulatory Framework and Public Acceptance

Technical Topic B: Barrier and System Performances within a Safety Case: their Functioning and Evolution with Time

Technical Topic C: The Role and Limitations of Modelling in Assessing Post-Closure Safety at Different Times

Technical Topic D: The Relative Value of Safety and Performance Indicators and Qualitative Arguments in Different Time frames

Each of the four working groups produced its own synthesis of discussions. These syntheses form Part B of the present document.

Discussions and presentations were, in general, limited to safety assessments covering the post-closure period, although the operational safety was also mentioned on several occasions (e.g. in [2]).

1.3   *Issues addressed and structure of this synthesis*

The main issues addressed by the workshop can be stated as follows.

a)   Why (if at all) do safety assessments need to address long timescales?

b)   How can the uncertainties arising from the need to address long timescales be dealt with?

c)   How can the limited predictability of the surface environment and of future human actions and habits be dealt with?

d)   How can the limits to the predictability of the geological environment be dealt with?

e)   What are the safety issues and timescales that concern the public?

f)   How do appropriate arguments for safety, or the relative weighting of different arguments, vary as a function of the timescales considered?

g)   How or to what extent should regulations guide the choice of safety arguments and their relative weighting?

The views expressed on these issues in the workshop are summarised in Sections 2 to 8 of this synthesis. The recommendations of the workshop are presented in Section 9.

2. Why do safety assessments need to address long timescales?

2.1 Summary of views

The long periods (often up to a million years or beyond) addressed in safety assessments, often as required by regulations, are a consequence of both ethical and technical issues.

Ethical principles dictate that consequences to future generations of measures to manage radioactive waste should never be greater than the levels of impact that are acceptable today and that the environment will continue to be protected in the future. The technical issues arise from:

- the long timescale over which the wastes present a potential hazard; and

- the fact that, in most situations, any releases to the surface environment only occurs in the distant future.

There was a general consensus that there is no real justification to prescribe a "hard" cut-off[1] to the period considered by safety assessments, although the nature of the arguments for safety may change over time.

2.2 Discussion

Repository safety studies are unusual in that they present statements about the evolution of natural and engineered systems over timescales are considerably in excess of those commonly considered in most engineering projects. Other types of environmental assessment commonly address periods of tens or occasionally hundreds of years. Safety assessments for deep geological repositories, however, address periods that are orders of magnitude longer than this, and are often required to do so by regulations. The workshop examined the reasons why this is the case.

Regulations generally set the minimum levels of protection that should be provided by repositories. Most regulations do not prescribe a cut-off time for the period to be addressed in safety assessments (see, e.g., [3,4,5]), although the ways in which protection is quantified may vary as a function of time. In several countries there is a requirement to assess impacts up to a calculated maximum of dose or risk, regardless of the time at which it occurs. Regulations in Switzerland state that doses and risks "shall at no time" exceed specified values. In the United Kingdom;

"The timescales over which assessment results should be presented is a matter for the developer to consider and justify as adequate for the wastes and disposal facility concerned" [5].

Some regulations suggest or prescribe timescales over which criteria based on dose or risk are applicable (see, e.g. [2]). This does not, however, imply that safety at longer times is of no concern, but rather, as discussed in later sections, that different arguments for safety may be more appropriate.

---

1. A specified time following which no arguments for safety need to be presented.

11

The workshop (see, e.g. [6]), and in particular Working Group A, reviewed a range of ethical and technical issues relevant to the timescales addressed by safety assessments. It is an ethical requirement that the measures adopted to manage radioactive waste should provide protection of both humans and the environment now and in the future. Indeed, it is one of the Principles set out in the IAEA Safety Fundamentals document [7] that "Radioactive waste shall be managed in such a way that predicted impacts on health of future generations will not be greater than relevant levels of impact that are acceptable today". Thus, although the minimum level of protection that a repository is required to provide can be based on ethical considerations, they give no real justification to prescribe a "hard" cut-off to the period considered by safety assessments, although the nature of the arguments for safety may change over time.

Disposal in suitably designed deep geological repositories is the strategy adopted to meet the ethical requirement to protect humans and the environment now and in the future. The strategy is, for an initial period at least, one of "concentrate and contain". Recent ICRP recommendations [8], however, point out that the "concentrate and contain" strategy leads to a concentration of the hazard and that eventually some release of radionuclides to the environment is virtually inevitable. Figure 2 in [3], for example, shows that, although there is a considerable reduction in the radiotoxicity of long-lived wastes over a million year period, beyond this time the radiotoxicity remains relatively constant, until one considers times up to 10^{10} years for which clearly no meaningful statement about containment can be made (it is pointed out in [3] that 10^{10} years is roughly equal to the expected remaining lifetime of our Sun). At later times, the siting and design of repositories must ensure that releases do not give rise to potentially harmful concentrations of radiation in the environment.

The long timescales addressed by safety assessments reflect in part the long timescale over which the wastes present a potential hazard. It should be noted, however, that toxic chemical wastes may remain toxic essentially for all times, and yet safety assessments for disposal facilities for these wastes span relatively short periods. It is probably that fact that radioactive isotopes have finite half-lives that originally gave rise to this difference in approach, and it is the half-lives of radioactive materials that seem to have set the timescales for safety assessments of radioactive waste repositories, even though the chemical toxicity of wastes may persist indefinitely into the future. Long timescales also reflect the fact that, in most situations, any releases to the surface environment only occur in the distant future. Indeed, according to [9], the fact that long timescales need to be addressed in safety assessments that aim to assess the maximum radiological consequences of a repository is, in itself, some indication of the power of deep geological disposal as a waste management solution.

3. How can the uncertainties arising from the need to address long timescales be dealt with?

3.1 Summary of views

Repositories are sited and designed where possible to avoid, or at least minimise, detrimental phenomena and uncertainties with the potential to undermine safety. The intrinsically favourable properties of the chosen site and design, including geological stability, should be emphasised in the safety case. Some uncertainties are inevitable and generally increase with the timescale considered. Valid arguments for safety can be made in spite of such uncertainties, but the limits of predictability of the evolution of the system must be clearly acknowledged in safety cases and should also be reflected in regulations. Techniques exist to categorise uncertainties, e.g. as scenario, model and parameter uncertainties, and to deal with these uncertainties in safety assessments. For uncertainties that can be quantified or bounded, these include the conservative approach and / or probabilistic techniques.

Alternative or complementary approaches for uncertainties that are difficult or impossible to bound are discussed further in Sections 4 and 5.

3.2   *Discussion*

The types of argument for safety that can be made were discussed at length by Working Groups B and C, and are closely tied to the use of safety and performance indicators discussed by Working Group D.

It is important to stress that deep geological repositories are sited and designed with long-term safety as a primary consideration. Detrimental phenomena and uncertainties with the potential to undermine safety are, where possible, avoided or at least minimised [10]. Thus, repositories may be sited and designed:

- to avoid or minimise complex interactions between engineered and geological materials;

- to minimise the possibility of rapid or sudden fundamental changes in the geological or geochemical environment, as the result of, for example, major tectonic movements or volcanic events, through the selection of a stable site [23], [11];

- to ensure that conditions deep underground are largely decoupled from events and processes occurring near the surface, including climate change; and

- to minimise the possibility of inadvertent human intrusion by avoiding sites with natural resources that might attract exploratory drilling.

The properties of engineered materials used in repositories are, in general, well understood, with a range of evidence, including observations from nature, supporting predictions of their behaviour and their long-term stability under expected conditions. The geological setting is also selected, at least in part, for its long-term stability. It was, however, noted by Working Group B that the concept of geological stability does not imply that steady-state conditions prevail in the geological environment, but rather, that its key safety-relevant properties (e.g. mechanical stability, low hydraulic conductivity, favourable geochemical conditions) are insensitive to man-made or natural perturbations.

The effects of remaining uncertainties can, to some extent, also be mitigated by the choice of site and design. The multi-barrier concept is generally employed, whereby the performance of the repository as a whole does not exclusively depend on a single feature or process. As illustrated, for example, in Figure 4 in [3] and Figure 1 in [12], if one feature or process contributing to safety is less effective than expected, or becomes ineffective earlier than expected, then other positive features or processes, termed "latent safety functions" in [12], can to some extent take its place. A specific example of the way the effects of uncertainties can be mitigated by design is provided by the long-lived containers envisaged in most high-level waste repositories. These mitigate the effects of uncertainties associated with the complex and coupled thermal, hydraulic, mechanical and chemical processes that occur during the transient phase following repository closure. If the containers remain intact throughout the duration of the transient phase, then provided the characteristics of the system after this phase can be well predicted, these uncertainties have limited implications for safety.

Since the safety of a well-sited and designed repository depends primarily on the favourable characteristics of the engineered barrier system (EBS) and host rock, including their predictability over a prolonged period, these characteristics need to be stressed in safety cases. This can be achieved by highlighting relevant *in situ* observations and measurements, examples being natural isotope profiles in some argillaceous rocks indicating a migration regime driven mainly by diffusion (as presented in [3]), and groundwater ages. Thermodynamic, kinetic, mass balance and palaeohydrogeological arguments can play a role here [e.g. 10], as can arguments for the feasibility, in principle, of safe geological disposal, for which relevant evidence is the existence of natural analogues and, in particular, natural uranium deposits.

Solely citing the favourable characteristics of the EBS and the host rock cannot however, make a safety case. All national regulations require some evaluations to be made of safety indicators that can be directly compared to specified yardsticks or reference values, and especially dose and risk. These evaluations have to consider the evolution of the disposal system (and require assumptions about surface environmental processes and radiological exposure modes), and, even for a well-chosen site and design, are inevitably subject to uncertainties. These uncertainties generally increase with time and do so at rates that vary significantly between different parts of the system. Eventually, but at very different times for different parts of the system, uncertainties are so large that predictions regarding their evolution cannot meaningfully be made. Figure 1, which is adapted from [3], illustrates that, at least for a well-chosen site, the evolution of the broad characteristics of the EBS and the host rock are reasonably predictable over a prolonged period (10^5 to 10^6 years, say, in the case of the host rock). There are uncertainties affecting the EBS and the host rock over shorter timescales, but these can, in general, at least be bounded with some confidence. The patterns of groundwater flow (the hydrogeological system), in particular near the surface, can be affected by climate change[2] and are thus somewhat less predictable. Surface environmental processes and radiological exposure modes are not generally considered to be parts of a deep geological repository system, but are relevant for evaluating dose and risk. These are less predictable still, being affected by ecological change, human activities and individual habits, which are highly uncertain, even on a timescale of a few years.

Over short enough timescales, many uncertainties can be bounded with reasonable confidence on the basis of current scientific understanding (as illustrated schematically by the darker shaded areas in Figure 1). Techniques that can be used to take such uncertainties into account, for example, include:

- the use of conservatively selected parameter values and conservative assumptions, that ensure that models used to assess the radiological consequences of the repository err on the side of pessimism;

- the use of probabilistic techniques, or a range of individually performed deterministic calculations, in order to ensure that the range of possibilities is fully explored in the calculations of radiological consequences.

It was noted by Working Group B that a conservative treatment may be acceptable, desirable or even required when demonstrating compliance with regulatory criteria, whereas a more realistic treatment is required for optimisation purposes and for the testing of models and databases.

Over longer timescales, depending on the part of the system under consideration (Figure 1), uncertainties generally increase and become more difficult to bound. The techniques mentioned above then become more difficult to apply, since the range of possibilities can become virtually unlimited. The implications for assumptions regarding future human behaviour and the evolution of the surface

---

2. With respect to climate change, [13] discusses recent work in the mathematical modelling of astronomical variations which indicated that the timing of glaciations can be predicted with a reasonable degree of certainty, although the magnitude of such events was subject to more uncertainty.

environment are discussed in Section 4. The need (if any) to make statements about the evolution of the EBS and the host rock over timescales where such statements become speculative, and how such statements can be supported, is discussed in Section 5.

Figure 1. **Schematic illustration of the limits of predictability of various aspects of a geological disposal system (adapted from [3]; note that actual timescales are site- and design-specific)**

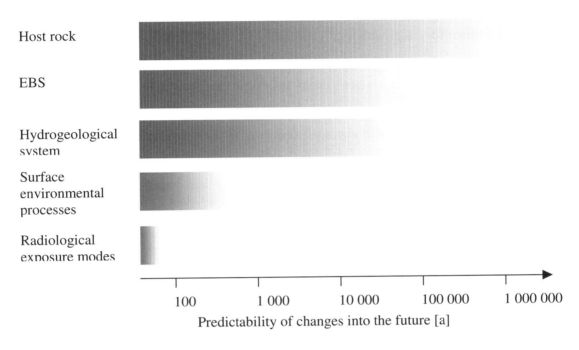

4. How can the limited predictability of the environment and of future human actions and habits be dealt with?

4.1 Summary of views

There is a consensus view that a stylised approach is appropriate for dealing with the very limited predictability of the surface environment and of future human actions. The stylised situations considered in this approach can, where necessary, consider a range of conditions, including different representative climate states. The issue was considered by the workshop to be effectively "solved", although it was stressed by several participants that the meaning of evaluated dose and risk as *indicators of safety*, rather than precise measures of expected consequences, needs to be stressed in the presentation of safety assessment results. The use of safety indicators complementary to dose and risk (e.g.: radiotoxicity fluxes from the geosphere) can help circumvent the limited predictability of the surface environment, provided accepted reference values or yardsticks for comparison can be derived.

4.2 Discussion

Many significant and largely unpredictable changes may occur that greatly modify the characteristics of society, human habits and the surface environment over timescales of a few tens to a few hundreds of years. Attempts to place bounds on the possible characteristics of the surface environment and the nature of future human actions over longer timescales are generally highly speculative.[3] According to the ICRP, dose can provide a measure of potential impact on human health for up to a few hundred years after repository closure [14]. There is, however, consensus that dose and risk can still be used as quantities for testing against regulatory and design targets, as long as it is appreciated that estimated values are to be regarded only as indicators of safety.

The role and treatment of the surface environment in safety assessments has been discussed extensively in international fora, e.g. within an NEA *ad hoc* group [15], [16] and, especially, within the IAEA BIOMASS project [1], [17]. In those countries represented at the workshop, the approach taken to dealing with uncertainties in the characteristics of the surface environment and the nature of future human actions is to define a range of "credible illustrations" or "stylised situations", mainly based on current conditions, as well as hypothetical critical groups as a basis for calculations of dose. Most countries apply this approach into the distant future. Where this involves a period of more than a few thousand years, the stylised situations consider a range of possible conditions, which can include different representative climate states. In some systems, the treatment of the surface environment can effectively be decoupled from the more predictable deep underground environment. In [3], for example, the effects of uncertainties related to the biosphere are explored using stand-alone calculations. In others, understanding of the deep underground environment (in particular, the rate of groundwater flow) must be coupled to often less well-supported assumptions made regarding the evolution of the surface environment. In general, the human nutritional needs and metabolism are assumed to be similar to those of the present day, and speculation about advances in science and technology (e.g. the development of cures for radiation-induced cancers) is excluded.

In most countries, the definition of the assumptions underlying these stylised situations (climate states, exposure pathways, etc.) is considered to be either a matter for the implementer or a matter for discussion between the implementer and the regulator. The Finnish regulator takes a somewhat different view [2]. Here, it is the regulator that defines exposure pathways, and the disruptive events (including those due to future human actions) that must, as a minimum, be considered by the implementer over the first few thousand years. Furthermore, beyond about ten thousand years (the "era of extreme climate changes", see Section 8.2), the regulator in Finland considers that the nature of the surface environment is so uncertain that it is more prudent to base the radiation protection criteria on constraints for release rates of radionuclides from geosphere to biosphere (geo-bio flux constraints), rather than on dose or risk constraints. The derivation of such constraints also involves, in effect, the making of some stylised assumptions regarding the surface environment and exposure pathways. In Finland, it is the regulator that has taken upon itself the burden of defining such assumptions.

5. How can the limits to the predictability of the geological environment be dealt with?

5.1 *Summary of views*

Well-supported statements regarding safety over a prolonged period can be made provided a repository is well designed and a suitable, geologically stable site is selected. A safety case may, in addition, make some statements regarding safety, even at very distant times when the evolution of the geological environment is less predictable. A less rigorous assessment of radiological consequences is likely to be adequate for these times, on account of the strongly decreased radiological toxicity of the

---

3. This is the reason why reliance on institutional control for disposal is limited to a few hundred years [14].

waste. Comparisons with natural systems can provide a basis for these statements. A stylised approach might also be appropriate at these times, but the acceptability of such an approach from the regulatory point of view requires clarification.

5.2   *Discussion*

For a well-chosen, geologically stable site and a well-designed repository, an adequate scientific database can, in principle, be obtained and used as a basis for statements on the future evolution of the repository and its environment (EBS and host rock) over a prolonged period. Indeed, the predictability of the geological environment for long times provides a key element of the safety case for all deep geological repositories. The length of the period over which well-supported statements can be made varies widely according to the system under consideration and, in particular, according to the degree to which the deep underground environment is isolated from changes closer to the surface. In the case of the Opalinus Clay in Switzerland [3], for example, observations from nature strongly suggest that diffusion has played a dominant role in controlling porewater compositions in this formation for more than a million years and, by extrapolation, is expected to control the movement of any radionuclides released into those porewaters from a repository over a similar timescale. The question that then arises is what statements, if any, can or should be made about safety beyond this period?

The participants of Working Group A felt that there is no objective reason to ignore the possibility that a repository might eventually be exposed at the surface by e.g. uplift and erosion, and the questions that should be explored are then "when could this happen?" and "what can be said about such a situation in terms of safety?".

There was general agreement that less weight should be placed on the evaluation of dose and / or risk in the distant future, due to the mounting uncertainties associated with these evaluations. This approach was seen to be acceptable due to the strongly decreased radiological toxicity of the waste. Some participants felt that there was little point in continuing dose and risk calculations beyond the period over which geological stability can be assured. Others felt that such calculations still have value, provided their limitations are explained [11]. The suggestion was also raised that, beyond the period over which geological stability can be assured, a stylised approach would be appropriate, similar to that adopted for uncertainties in the evolution of the surface environment and future human actions, in order to avoid "pointless speculation", although the acceptability of such an approach from the regulatory point of view would need clarification.

As the uncertainties associated with calculations of dose and risk increase, other complementary safety or performance indicators that are less affected by these uncertainties may take a more prominent role, as discussed by Working Group D and in [18] and [19] (and as required by regulations in Finland [2]). One such indicator that participants felt could (and does) contribute to the safety case at times when even the predictability of the geological environment is limited is that of the radiological toxicity of the emplaced waste. This can be compared with natural systems, such as uranium ores, and can give an additional argument for safety at times when less (or even zero) weight is attached to the results of dose and risk calculations.

Hedin [20] and Nagra [3] gave examples of the use of radiological toxicity as an indicator.

- Hedin [20] showed that after about 100 000 years the radiological toxicity of one tonne of Swedish spent fuel is on a par with the radiological toxicity of the natural uranium from which it was derived. This crossover point suggests a return to radiation levels found in natural materials. It thus defines an easily communicated timescale for a natural assessment endpoint [6], and is proposed as a timescale for which a repository must function in the SR97 safety assessment.

- Nagra [3], in its assessment of a repository for spent fuel, reprocessed high-level waste and long-lived intermediate-level waste in Opalinus Clay, compared the RTI[4] of the waste with that of the natural radionuclides contained in 1 km^3 of Opalinus Clay and with that of a volume of natural uranium ore corresponding to the volume of the SF / HLW emplacement tunnels, and used this to argue that one million years is the period of principal concern.

Working Group A cautioned that such comparisons must be used with care. Different types of plots can lead to different crossover times, thus potentially weakening the confidence in this type of argument, and activity or toxicity curves alone have limited meaning from the point of view of risk and safety, since, for example, the mobility of radionuclides is not taken into account. The participants, however, acknowledged the potential value of this approach, and Working Group A suggested that the comparison with natural radioactive systems provides a basis to be explored for future regulatory guidance.

6.   What are the safety issues and timescales that concern the public?

6.1   Summary of views

Some participants expressed the view that, in the light of feedback from the public and other stakeholders, safety assessments have often placed too little emphasis on the initial period of several hundreds to a few thousands of years. It was suggested that dialogue with the public, and indeed the wider scientific community, might lead to less emphasis being placed on very long timescales in a safety case, and more emphasis on the shorter timescales that the public can better envisage. The workshop noted, however, that the public is heterogeneous in its concerns and public interest in longer timescales should not be discounted, particularly with respect to the impact of scenarios such as glaciation, which the public can readily envisage. Safety and performance indicators complementary to dose and risk and arguments based on the stability and predictability of the geological environment, provide valuable means of communicating the safety case to the public.

6.2   Discussion

The workshop noted an increasing recognition that the presentation of the safety case needs to be tailored to address the concerns of the intended audience. In [9], a few representative audiences and their possible interpretations of safety and timescale are considered, including political decision-makers, regulators, the scientific and technical community and the "public". The workshop expressed a view that greater dialogue with the public is required in order to identify public concerns. To this end, Nirex has developed a methodology that divides the period covered by safety assessment into a set of "nested time frames" [21], which is designed, in part, to encourage public dialogue. These nested time frames can provide a tool to answer "what if?" type questions raised by the public.

Working Group A and others at the workshop recognised the heterogeneity of the public and its concerns, and noted that the interest of the public in longer timescales should not be discounted. There was, however, consensus that the long timescales addressed in safety assessments are perhaps difficult for the public as a whole to digest, and statements made for such periods difficult to accept. They are orders of magnitude greater than the timescales of direct human experience, and may appear to stretch scientific knowledge significantly, particularly given the occasional failure to predict (sometimes with catastrophic consequences) the behaviour of engineered structures in the much shorter term. It was suggested that many members of the public would probably consider 1 000 years as a very long timescale and times beyond that might not be seen as sensible or credible. This is the motivation for the increased emphasis on the first thousand years by the Swedish regulator [22]. The sceptical attitude towards long-term prediction of the evolution of the engineered barriers is noted particularly in [9,6,.23].

In recent years, the importance of using a wide range of arguments as a means of explaining the safety case to a wider range of audiences has grown. Safety and performance indicators complementary to dose and risk have potential value in explaining safety arguments to non-technical audiences by providing comparisons of potential impacts with, say, current natural conditions. Furthermore, where high integrity canisters provide an initial period of complete containment that is expected to last for thousands of years or more, complementary indicators can be used to emphasise the fact that zero release to the host rock (and a substantial decay of decreased radiological toxicity) is expected over this prolonged period. Other arguments for long-term safety that may be convincing to the public, as well as the wider scientific community, include those based on the stability and predictability of the geological environment. The need for complementary presentational methods (in addition, for example, to log-log graphs of dose or risk vs. time) was noted, so that the meaning of safety assessment results is made clear to less specialised audiences.

### 7.	How do appropriate arguments for safety, or the relative weighting of different arguments, vary as a function of the timescale considered?

#### 7.1	Summary of views

Both hazards associated with radioactive waste and uncertainties associated with the evolution of a repository vary with time, and a safety case has to make the best use of the arguments for safety that are available at any given time. Due to a better appreciation of the limitations of models used to evaluate dose and risk, and in response to the need to communicate safety assessment findings to a wider audience, there is a trend towards employing multiple lines of reasoning in safety cases, with an emphasis on different indicators of performance and safety at different timescales, some of which are more qualitative in nature. In general, the safety case for the initial period, when radionuclides are contained predominantly or entirely within the repository itself, should emphasise arguments that give confidence in a high degree of (or even complete) containment. At times when containment in the repository can no longer be relied upon to prevent radionuclide releases, the safety case must rely increasingly on evaluations of dose and risk, with a range of arguments supporting model assumptions and parameter values. There were differences in views as to how long dose and risk evaluations should be continued. There was, however, consensus that arguments related to the lower radiological toxicity of the waste are more appropriate, or should carry more weight, at later times, when reliable predictions of the evolution of the geological environment can no longer be made.

#### 7.2	Discussion

The presentations and discussions at the workshop demonstrated a trend away from reliance on calculated dose or risk as the sole indicators of safety, and towards employing multiple lines of reasoning, with an emphasis on different indicators of performance and safety at different timescales, some of which are more qualitative in nature [11,24]. This is in part due to an increased awareness that more direct evidence for the intrinsic qualities of the site and design can provide convincing arguments for safety that may be easier to communicate to a wide, non-specialist audience. In part, it is also due to recognition of the need to acknowledge the limitations of the models used to evaluate dose and risk, particularly at very distant times in the future.

Factors determining the most appropriate arguments for safety or the relative weighting of different arguments, at different times include:

- the concerns of the intended audience for the safety case, including the public;

- the types of phenomena occurring within different time periods that affect the fate of radionuclides (including the decrease in radiological toxicity with time); and

- the degree of uncertainty associated with these phenomena.

Ethical considerations can also play a role. Working Group A discussed at length how regulatory requirements should balance risks over short and long times and hence address the question of how the regulator may judge compliance at different times, taking into account ethics, science, public opinion and public acceptance.

In view of the different phenomena and uncertainties that characterise different periods in the evolution of a repository and its environment, many programmes and some regulations divide the future into a number of discrete intervals or "time frames", with different arguments, or weightings of arguments, developed for each [4,12,24]. The "nested time frames" used by Nirex have already been mentioned in Section 6. Time frames can provide a useful framework for internal discussions among experts within an implementing organisation, between implementers and regulators and between implementers, regulators and the public. Working Groups B and C, for example, structured their discussions by dividing the post-closure period into a number of generic time frames. Similarities were noted in the reasoning used to delineate time frames, but the time frames themselves are programme-, concept- and site- specific.

Following closure of the repository,[4] there is generally a period during which complete containment of radionuclides is expected. Following from the suggestion in Section 6 that the public and other stakeholders are primarily interested in the initial period of several hundreds to a few thousands of years, it was suggested that the safety case for this period should emphasise clearly the arguments that give confidence in a high degree of (or even complete) containment. In the case of spent fuel and high-level waste, in view of the high potential hazard, this period is of particular importance. The period of complete containment is likely to coincide, at least to some extent, with a phase of relatively complex transient phenomena, including resaturation of the repository and its surroundings. If complete containment can be assured during the transient phase, this can reduce the need to model these phenomena in detail, although the implications of transient phenomena on the longer-term characteristics of the disposal system must be considered.

At times when containment in the repository can no longer be relied upon to prevent radionuclide releases, the safety case must rely increasingly on evaluations of dose and risk, with a range of arguments supporting model assumptions and parameter values. In view of the fact, however, that the uncertainties associated with the evolution of the repository and its environment generally

---

4. In many national programmes, there are proposals for an extended period of monitored, retrievable underground storage, during which the repository may be kept open and unsaturated.

increase with time, while the hazard associated with the waste decreases, detailed calculations, including evaluations of dose and risk, should eventually carry less weight than arguments related to decreasing radiological toxicity and the remaining containment or isolation capability of the disposal system. The limitations of use of arguments based on radiological toxicity of the emplaced waste as a safety indicator were, however, noted.

8. How or to what extent should regulations guide the choice of safety arguments and their relative weighting?

8.1 Summary of views

Regulations are increasingly providing guidance as to the acceptability of, or the weight that can be attached to, arguments that may complement or even replace dose and risk calculations at later times.[5] If calculations of dose or risk are required even beyond the period when the stability of the geological environment can be assured, guidance should be given on the range of possibilities or scenarios that these calculations need to explore, and to the weight that should be attached to them. Regulations should not place undue demands on the modelling capabilities required to evaluate specific safety or performance indicators.

8.2 Discussion

As discussed in earlier sections, regulations do not generally impose a "cut-off" time for the period to be addressed in safety assessments. Many, however, acknowledge the changing nature of the arguments that are appropriate at different times, or within different time frames.

Regulations in the United Kingdom imply a time frame in which detailed calculations of risk and dose are appropriate and a more distant time frame in which simpler scoping calculations and supporting qualitative information are more appropriate, when the validity of models of radionuclide release and transport becomes questionable [9]. In Sweden, a risk limit is set without any time limitation. Quantitative analyses of the impact on human health and the environment are, however, required only for the first thousand years, whereas, in the subsequent period, the requirements are less well defined and the objective becomes to assess the protective capability of the repository system based on various possible scenarios [22]. The Finnish regulator is more precise in its requirements, explicitly defining the following post-closure time frames for safety assessments for spent fuel [2]:

- the "environmentally predictable future" (several thousand years), during which conservative estimates of dose must be made;

- the "era of extreme climate changes" (beyond about ten thousand years) when periods of permafrost and glaciation are expected, radiation protection criteria are based on geo-bio flux constraints; and

- the "farthest future" (beyond about two hundred thousand years), when the activity in spent fuel becomes less than that in the natural uranium from which the fuel was

---

5. Some members of Working Group A thought that no weight should be placed on the evaluation of dose and risk beyond about 100 000 years, which is the crossover point, used in [21] (see section 5.2). Other members, however, preferred not to define a hard "cut off" time for the meaningfulness of such calculations.

21

fabricated, for which no rigorous quantitative safety assessments are required and statements regarding safety can be based on more qualitative considerations.

In [6], in order to balance ethical and technical considerations and public concerns, a series of time-graded containment objectives is suggested with two target times. It is suggested that the initial period of 500 years corresponds to the period of greatest public concern. For this period the objective of total containment is proposed, at least for spent fuel and reprocessed high-level waste. In the time period up to 100 000 years – the end point roughly corresponding to the crossover point on activity curves – a dose constraint derived from natural background radiation levels is prescribed. Beyond some 100 000 years the system is approaching a natural one and the proposed objective is that the eventual redistribution of the residual activity by natural processes remains indistinguishable from natural regional variations in radiation levels.

9.   Conclusions and recommendations

9.1   Conclusions

Conclusions regarding the main issues addressed by the workshop are as follows.

(i)   Why do safety assessments need to address long timescales?

Safety assessments need to address long timescales due to the inherent characteristics of the waste and of geological disposal and because of ethical considerations. These characteristics and considerations are (i), the long period over which the radiological toxicity of the waste persists, (ii), the fact that, for a well-sited and well-designed system, any releases of radionuclides are expected to occur very far in the future (if at all), and (iii), the ethical principles that require the same level of protection for humans and the environment in the future as that which is applicable today. The workshop contrasted the long timescales addressed in safety assessments for radioactive waste repositories with the shorter timescales addressed in safety studies of facilities for other wastes. It was noted that it is probably that fact that radioactive isotopes have finite half-lives that originally gave rise to this difference in approach, and it is the half-lives of radioactive materials that seem to have set the timescales for safety assessments of radioactive waste repositories, even though the chemical toxicity of wastes may persist indefinitely into the future.

(ii)   How can the uncertainties arising from the need to address long timescales be dealt with?

Although, where possible, uncertainties are avoided, or their effects mitigated, by the choice of appropriate site and design, there are significant uncertainties related to the future evolution of the system that must be taken into account in the assessment of the safety of a repository system. The importance of such uncertainties generally increases with time. Methods, however, exist whereby a safety can be evaluated in the presence of uncertainties, including, for example, probabilistic methods, use of conservatism and the use of stylised approaches. More generally, multiple lines of reasoning, with an emphasis on different types of argument and different indicators of performance and safety at different timescales, some of which are more qualitative in nature, can be used in the building of a safety case. In order to maintain credibility within the scientific community as well as with other stakeholders, it is important to acknowledge the limits of predictability of the system in both regulations and in safety cases.

(iii)   How can the limited predictability of the surface environment and of future human actions and habits be dealt with?

Stylised approaches, such as, for example, reference biospheres, are appropriate for dealing with the limited predictability of the surface environment and of future human actions, and the issue was considered by the workshop to be effectively "solved". In addition, the use of safety indicators complementary to dose and risk (e.g.: radiotoxicity fluxes from the geosphere) can help circumvent the limited predictability of the surface environment, if accepted reference values or yardsticks for comparison can be derived. In some countries, the choice of stylised approaches is considered a matter for the implementer. Elsewhere, it is a matter for the regulator, or for dialogue between the implementer and the regulator. In view of the stylised approaches used for the surface environment and of future human actions, calculated doses and risks should not be interpreted as actual measures of health detriments and risks to future individuals, but rather as illustrations based on agreed sets of assumptions.

(iv)   How can the limits to the predictability of the geological environment be dealt with?

In spite of the inevitable limits to the predictability of any geological environment, the methods of geological sciences are considered adequate to show, for a suitably chosen site, that properties favourable to the long-term safety of a repository will be maintained for a sufficient time for radiological toxicity to decrease substantially. Thus, as the uncertainties in the results of calculations of dose and risk increase due to increasing uncertainties in the evolution of the geological environment, the hazard associated with the radiological toxicity of the waste declines.

There is consensus that weighting attached to calculations of dose and risk should decrease with time as a consequence of increasing uncertainty, and that, at later times, more weight can be placed on other performance and safety indicators (e.g. radiological toxicity of the waste), with yardsticks drawn e.g. from natural systems.[6] The use of a stylised approach for evaluating dose and risk at times when geological stability can no longer be assured was suggested, but needs further clarification. It was also felt that the issues of geological stability and predictability require more clarification (see "possible future work", below).

(v)   What are the safety issues and timescales that concern the public?

The first few hundred years following emplacement of the waste is probably the period of highest concern to many members of the public and should be emphasised to a greater degree in safety cases addressed to the public. Generally, zero release is expected during this period, and this point could be better argued in safety cases. The public is, however, heterogeneous and public interest in longer timescales should not be discounted, particularly with respect to the impact of scenarios such as glaciation, which the public can readily envisage. Safety cases addressed to the public need to be presented in an understandable manner, and complementary safety and performance indicators, as well as a range of presentation techniques, were suggested as having a role in this respect. As mentioned above, it was noted that an acknowledgement of the limits of predictability of the system is important for credibility in the eyes of the public and of other stakeholders.[7]

(vi)   How do appropriate arguments for safety or the relative weighting of different arguments, vary as a function of the timescale considered?

---

6.   Working Group D recommended that the compilation work on natural concentrations and fluxes co-ordinated by the IAEA [7] should be continued.

7.   Here, the term stakeholder is a convenient label for any actor, institution, group or individual with a role to play in the process (see proceedings of the FSC workshop, Paris, 2000).

Both hazards associated with radioactive waste and uncertainties associated with the evolution of a repository vary with time, and a safety case has to make the best use of the arguments for safety that are available at any given time. Many programmes and some regulations define time frames, in each of which different types of argument for safety are provided, or different weightings are assigned to certain types of argument. Similarities were noted in the reasoning used to delineate time frames, but the time frames themselves are programme-, concept- and site-specific. In many national programmes, there are proposals for an extended period of monitored, retrievable underground storage, during which the repository may be kept open and unsaturated. Following closure of the repository, there is generally a period during which complete containment of radionuclides is expected and safety cases concentrate on the arguments that support this expectation. In the case of spent fuel and high-level waste, in view of the high potential hazard, this initial period is of particular importance. The initial period is likely to coincide, at least to some extent, with a phase of relatively complex transient phenomena, including resaturation of the repository and its surroundings. If complete containment can be assured during the transient phase, this can reduce the need to model these phenomena in detail, although the implications of transient phenomena on the longer-term characteristics of the disposal system must be considered. As soon as complete containment can no longer be assured, dose and risk become the primary indicators of safety, with a range of arguments supporting the model assumptions and parameter values used to evaluate these indicators. At later time, for example at times beyond limits of predictability of geological setting, there is a trend towards safety cases concentrating on the decreased radiological toxicity of the waste. The limitations of use of arguments based on radiological toxicity of the emplaced waste as a safety indicator were, however, noted at the workshop.

(vii)   How or to what extent should regulations guide the choice of safety arguments and their relative weighting?

Regulations are increasingly taking into account the full range of arguments for safety that is available, in addition to calculations of dose and risk, and providing guidance regarding their use. Although there are no scientific or ethical arguments that justify imposing a definite limit in regulations to the period to be addressed by safety assessment, regulations should not place undue demands on the modelling capabilities required to evaluate specific safety or performance indicators. With regard to the evaluation of dose and risk, some cut-off time will, in practice, inevitably be applied in safety assessments. This may be dictated by regulations, or it may be the result of a decision of the implementer or discussions between regulators and implementers. Nevertheless, as suggested above, other, possibly more qualitative arguments for safety at times beyond any such cut-off should be provided. There is thus a preference for a shift in the weights given to the different arguments and safety and performance indicators when judging compliance on longer timescales. This shift in weight should reflect the increasing uncertainties over time, and also the decreasing radiological toxicity of the waste. There was some disagreement as to how exactly weights should be assigned to different arguments at different times, but it was agreed that regulations could usefully provide further guidance and clarification on issues related to the weighting of arguments, including how exactly weighting should be defined (see "possible future work", below).

9.2   Recommendations

9.2.1   Presentation of the safety case

Safety assessment results, their purpose and their interpretation need to be explained carefully in presenting a safety case in order to maintain credibility with various audiences [2]. The presentation of a safety case for the period of a few hundred years following emplacement of the waste may deserve further attention, with greater emphasis being placed on this period in documents aimed at the public. Furthermore, irrespective of timescale, such documents should highlight arguments that the public might find persuasive, including arguments based on direct observation of the intrinsic quality of the system, and safety and performance indicators that allow the performance of the disposal system to be placed in perspective with natural phenomena (e.g. natural radionuclide fluxes).

Even for a technical audience, the presentation of dose or risk as a function of time is, on its own, not an effective way to convey the message that deep geological repositories provide an appropriate level of safety. It can be useful to complement graphs of dose or risk as a function of time by additional graphs or tables that more directly illustrate the performance of the different repository barriers. For example, illustrating the decay of the overall toxicity of the waste as a function of time can usefully illustrate the rationale behind the "concentrate" and "contain" strategy, i.e. isolation and containment of the waste while radioactive decay reduces the associated hazard.

9.2.2   *Possible future work*

The following are recommendations for future work that could be helpful to clarify issues and improve the approaches for dealing with them. They are presented under the four subject areas of:

- geological stability and predictability;

- regulatory guidance;

- identifying and addressing public concerns; and

- the definition of time frames.

Geological stability and predictability

The issue of geological stability, and how the limits to the predictability of the geological environment can be considered, is a concern to many programmes. This issue needs further clarification with respect to its implications for the safety case. Specific points to be addressed could include:

- How should geological "stability" be defined (it does not necessarily imply that the geological system is in a steady state)?

- What degree of geological environment stability can be expected for a well-sited repository?

- To what extent do arguments for safety need to be made beyond the timescale over which predictions of the evolution of the geological environment can reasonably be made?

- Is a "stylised approach", similar to that adopted for the evolution of the surface environment and future human actions, appropriate to deal with these very long timescales?

Regulatory guidance

While regulations should not become too prescriptive, they could usefully provide further guidance and clarification in areas related to timescales, including:

- The role of safety assessment, i.e. that it is to assess isolation potential, rather than to make predictions of the future.

- The definition of time frames and the meaning that should be attached to calculated doses and risks in different time frames.

- The acceptability and role in a safety case of multiple lines of reasoning, including safety and performance indicators other than dose and risk and qualitative information, and how safety arguments based on these are to be weighted in compliance judgements (including what exactly should be understood by "weight").

- The acceptability of calculated doses exceeding regulatory constraints over long timescales or for specific scenarios (e.g. human intrusion scenarios).

- The acceptability of complementary arguments based on, for example, the favourable characteristics of the site and natural analogues, at times in the distant future when the applicability of the models used to calculate dose and risk is questionable.

- The acceptability of a "stylised approach" for long timescales.

Identifying and addressing public concerns

In order to be accepted as part of a licence application for a repository, a safety case must address the concerns of the wider public. Indeed, according to [29], decisions regarding whether, when and how to implement geological disposal will require thorough public examination and the involvement of all relevant stakeholders. It was noted that:

- The trend towards employing multiple lines of reasoning in safety cases, with an emphasis on different indicators of performance and safety at different timescales, is partly a response to the need to communicate safety assessment findings to a wider audience, including the public.

- Dialogue with the public, though widely recognised as desirable, can be difficult to achieve, and methods for achieving constructive dialogue could usefully be developed further. Such methods must account for the fact that the public is a heterogeneous group that will bring a wide range of personal values to any dialogue process, making consensus potentially difficult.

- Currently conflicting information on the interests of different segments of the public regarding, for example, timescales, levels and types of information and level of protection, need to be resolved.

- Dialogue may lead to a different emphasis on the timescales adopted in repository safety assessments, possibly with less emphasis on very long timescales, as mentioned above.

The definition of time frames

There is consensus that the definition of time frames is a useful concept. Further consideration could be given to the development of methods for structuring such time frames, based on the understanding of the evolution of a disposal system or on other considerations, such as the limits of scientific predictability, the decrease of radiological toxicity, human concerns and consensual agreement among stakeholders.

REFERENCES

References to papers in Part C: 2, 3, 4, 6, 9, 10, 11, 12, 13, 18, 19, 21, 22, 23, 24, 25

External references: 1, 5, 7, 8, 14, 15, 16, 17, 20, 26, 27, 28

[1]   Confidence in the Long-term Safety of Deep Geological Repositories. Its Development and Communication, available from the NEA, Paris, 1999.

[2]   Consideration of Timescales in the Finnish Safety Regulations for Spent Fuel Disposal, E. Ruokola.

[3]   Handling of Timescales in Safety Assessments: the Swiss Perspective, J. Schneider, P. Zuidema & P. Smith.

[4]   Timescales in the long-term Safety Assessment of the Morsleben Repository, Germany, M. Ranft & J. Wollrath.

[5]   Radiation Protection Recommendations as Applied to the Disposal of Long-lived Solid Radioactive Waste, ICRP Publication No. 81, Pergamon Press, Oxford and New York, 2000.

[6]   Long Timescales, Low Risks Rational Containment Objectives that Account for Ethics, Resources, Feasibility and Public Expectations – some thoughts to provoke discussion, N. A. Chapman.

[7]   Spent Nuclear Fuel – How dangerous is it? A. Hedin, SKB report TR 97-13, Stockholm, 1997.

[8]   The Principles of Radioactive Waste Management, IAEA Safety Series No. 111-F, IAEA, Vienna, 1995.

[9]   Some Questions on the Use of Long Timescales for Radioactive Waste Disposal Safety Assessments, R. A. Yearsley & T. J. Sumerling.

[10]   Treatment of Barrier Evolution: The SKB Perspective, A. Hedin.

[11]   Handling of Timescales: Application of Safety Indicators, H. Umeki & P. Smith.

[12]   Fulfillment of the long-term safety functions by the different barriers during the main time frames after repository closure, P. De Preter & Ph. Lalieux.

[13]   Long Scale Astronomical Variations in our Solar System: Consequences for Future Ice Ages, K. G. Karlsson & S. Edvardsson.

[14]   Working group on biosphere analysis in repository assessments: The case for benchmark biosphere and de-coupling of biosphere and EBS/geosphere analyses, NEA PAAG document NEA/PAAG/DOC(98)6, 1998.

[15]   The role of the analysis of the biosphere and human behaviour in integrated performance assessments, OECD Nuclear Energy Agency PAAG document NEA/TWM/PAAG(99)5, 1999.

[16]   Long-term releases from solid waste disposal facilities: the Reference Biosphere concept. BIOMASS Theme 1 Working Document BIOMASS/TI/WD01, available from the IAEA, Vienna, 1999.

[17]   Alternative assessment contexts: implications for the development of reference biosphere's and biosphere modelling. BIOMASS Theme 1 Working Document BIOMASS/T1/ WD02, available from the IAEA, Vienna, 1999.

[18]   The SPIN Project: Safety and Performance Indicators in Different Time frames, R. Storck & D. A Becker.

[19]   IAEA Activities Related to Safety Indicators, Time frames and Reference Scenarios, B. Batandjieva, K. Hioki & P. Metcalf.

[20]   Radiation Protection Recommendations as Applied to the Disposal of Long lived Solid Radioactive Waste, ICRP Publication No. 81, Pergamon Press, Oxford and New York, 2000.

[21]   Handling timescales in post-closure safety assessments: A Nirex perspective, L. Bailey & A. Littleboy.

[22]   Safety in the First Thousand Years, M. Jensen.

[23]   Handling of Timescales In Safety Assessments of Geological Disposal: An IRSN-GRS Standpoint on the Possible Role of Regulatory Guidance, K.J. Rohlig & D. Gay.

[24]   Handling of Timescales and Related Safety Indicators, L. Griffault & E. Fillion.

[25]   Handling of Timescales: Application of Safety Indicators, H. Umeki.

[26]   Radioactive Substances Act 1993 - Disposal Facilities on Land for Low and Intermediate Level Radioactive Wastes: Guidance on Requirements for Authorisation: Environment Agency, Scottish Environment Protection Agency, and Department of the Environment for Northern Ireland, Bristol, January 1997.

[27]   Guidance on the definition of critical and other hypothetical exposed groups for solid radioactive waste disposal, Working Document BIOMASS/T1/WD03, available from the IAEA, Vienna.1999.

[28]   Biosphere System Identification and Justification. Working Document BIOMASS/T1/WD06, available from the IAEA, Vienna, 1999.

[29]   NEA, Stakeholder confidence and radioactive waste disposal, workshop proceedings, Paris, France, August 2000.

PART B

WORKING GROUPS CONTRIBUTIONS

Group A

THE DIFFERENT TIMESCALES *VERSUS* THE REGULATORY FRAMEWORK AND PUBLIC ACCEPTANCE

Chairperson: Johannes O.Vigfusson (HSK, Switzerland)
Rapporteur: Didier Gay (IRSN, France)

Members: Lucy Bailey (NIREX, UK); Neil Chapman (NAGRA, Switzerland)
Peter De Preter (NIRAS-ONDRAF, Belgium); Mikael Jensen (SSI, Sweden)
Ludger Lambers (GRS, Germany); Philippe Raimbault (DGSNR, France)
Esko Ruokola (STUK, Finland); Stig Wingefors (SKI, Sweden)
Roger Yearsley (Environment Agency, UK)

The questions and topics proposed to Working Group A for discussion fall into the following broad categories:

- current views and requirements in national regulations, notably with regards to the prescription of a cut-off time;

- ethics and public opinion;

- credibility and a need for long-term assessment.

1. Introductory presentations

Neil Chapman and Michael Jensen gave the initial input to the discussions within two introductory presentations. The full text of these presentations is found in Part C. Thus only the points most relevant for the further discussion in the group are highlighted here.

Neil Chapman – Long timescales, low risks: balancing ethics, resources, feasibility and public expectations

The author noted that the question of long timescales has connections with a series of facts and values that need to be balanced when providing guidance on this subject.

Values of concern can be found in ethical principles such as the principles of sustainability, precaution and chain of obligation. These principles indeed offer guidance on how to integrate consideration of the future into present decisions. Yet they also tend to give rise to conflicting interpretations notably with regards to the necessary trade-off between present day benefits and the reduction of hypothetical future risks. They thus lead to the difficult questions of how to allocate resources between current and future problems and how to apply sustainability regarding various time scales.

A scientific fact to be considered is provided by the decay of spent fuel activity with time. When plotted together with the natural activity contained in an equivalent uranium ore, a "crossover point" is reached after around 150 000 years. This crossover point suggests a return to radiation levels found in natural materials and thus defines an easily communicated time scale for a natural assessment endpoint.

Perception of time and presentation of risk over long timescales are other important aspects to be considered. It raises the following difficulties:

- An apparent gap seems to exist between times that matter to people and those invoked in geological disposal studies. For the layman, long-term typically means several decades, a few generations, whereas, for geological disposal specialists, long-term safety usually covers periods of several tens or hundreds of thousands of years. However, the possibility to restrict the time scope of concern through a prescribed time cut-off does not seem acceptable from an ethical point of view: it could be interpreted as a tacit discounting of future harm and therefore could be considered contrary to the principle of sustainability.

- Containment times referred to in assessments typically embrace durations that exceed most available historical references: as an example, container lifetimes might be expected to last around 5 000 years, which is longer than the whole of recorded history.

- Contrary to this, in current society there is a widespread feeling that predictions regarding the reliability of technical systems are usually proved wrong over periods as short as several decades, at least in the absence of adequate maintenance. Thus there is a sceptical attitude towards long-term prediction of engineered barriers.

Chapman suggested a possible way forward to cope with the challenges posed by distant time frames. It is based on giving emphasis to the comparison with the natural environment and on the definition of time graded containment objectives with two target times. An initial period of 500 years corresponds to the period of greatest public concern. For this period the objective of total containment is proposed. In the time period up to 100 000 years – the end point roughly corresponding to the crossover point on activity curves – a dose constraint derived from natural background radiation levels is prescribed. Beyond some 100 000 years, the system is approaching a natural one and the proposed objective is that the eventual redistribution of the residual activity by natural processes remains indistinguishable from natural regional variations.

Michael Jensen – Safety in the first thousand years

Swedish regulations stipulate that the annual risk of harmful effects after closure must not exceed 10^{-6} for a representative individual in the group exposed to the greatest risk. No time limit is specified for the application of this objective, and in fact the regulations provide little guidance on how to handle different timescales in the assessment of final management options. Although a quantitative limit is not prescribed, the regulations require the implementer to report collective doses over the first 10 000 years and to describe the protective capability of the repository after a potential human intrusion. The risk limitation is developed further in SSI FS 98:1. In this document a distinction is made between the thousand-year period following repository closure and the period beyond. Quantitative analyses of the impact on human health and the environment are required for the first thousand years. In the following period, the requirements are less well defined and the objective is to assess the protective capability of the repository system based on various possible scenarios. This reflects the intention to emphasise the importance of the first thousand-year period.

The importance given to the initial time period has several justifications. A time frame of several hundred years is compatible with what we consider an accepted societal horizon of prediction for engineered structures. It is also still relevant within our current concept of legal responsibility. The perspective and the references we commonly accept and use remain reasonable and meaningful in this period. Good assessment and good description of the initial evolution of the repository may also be a convincing argument and a bonus for enhancing confidence in the disposal project. It demonstrates that the system is adequately understood and thus may help strengthen the conviction that it is and will remain safe. Evaluation in the 1 000-year period also gives the opportunity to address issues that are closer to public concern and more concrete and accessible than considerations relative to a more distant future. This could contribute to bridging the gap that currently exists between public expectations and the type of information produced by implementers in their assessments.

The recommendation for a greater emphasis on the first 1 000-year period does not mean that consideration of the farther future should be abandoned. Rather, it means that a different treatment should be adopted for each period. In the first period, it is proposed to focus the assessment on the identification of the various possible events and situations likely to provoke short-term release and to discuss and assess their probabilities and their impacts. In the second period, beyond the first 1 000 years, the influence of various sources of uncertainty has to be acknowledged. This requires a more complicated safety assessment and tends to limit its potential value. In any case, in the far future the possibility that the repository may be significantly disturbed becomes a potentially reasonable scenario. For example, in 10 million years from now, about 100 successive glaciations will have eroded the site and possibly the repository itself. Therefore, the use of multiple lines of arguments (including qualitative ones) is recommended when addressing this time scale in the safety assessment.

2.   Summary of the discussions

Both introductory presentations and the ensuing discussions suggested that one key issue ultimately underlies any consideration about timescales and regulation: the issue of how to balance information relative to short- and distant-time risks and hence how to address the question of compliance judgement in different time frames. "Can we judge in the same way an impact occurring now and in 50 000 or 1 million years?" was thus a central question for discussion in this group. This issue was tackled from three different points of view: ethics, science and public acceptance.

Ethical viewpoint

The first spontaneous answer with regards the ethical point of view underlined that no discounting of future harm should be accepted. It was also rapidly acknowledged that although the basic principles of ethics are easily understood and accepted, it can be a difficult and controversial task to decide how they are adequately applied in practice. Depending on the interpretation and weight given to the various principles, it is easily possible to reach different conclusions and even justify opposite positions. In the case of geological disposal, there is an additional difficulty relative to the application of the principles to future situations. Thus, even though the ethical principles behind the requirement to offer equal protection to present and future generations (for example, the related principles of fairness and intergenerational equity) can be fully understood and approved, the concepts tend to have unclear meaning when applied to periods of time several tens of thousands or a million years distant in the future. The real level of protection that will result for people that may possibly live on the site at some time in the distant future is not known nor is it known what level of protection they may actually need.

Facing up to this apparent philosophical dead-end, participants suggested that the basic ethical principles define general objectives. Bearing this in mind, the aim is neither a direct transcription of ethical principles into the design objectives of disposal system, nor a demonstration of a literal compliance with them. Rather, the aim is to ensure that the option selected in practice is respectful of the principles and that the regulator (when defining the regulation) and the implementer (when running its project) effectively do their best to offer future generations the same level of protection as required today. In other words, ethics is regarded as a question of objectives within given capabilities, rather than solely a question of results obtained.

A strategy cannot be considered acceptable if it implies discounting of future harm. Thus the level of protection currently offered is also the reference for future protection. Considering the large uncertainties regarding the evolution of society and the biosphere, this lends justification to the use of current biosphere, technical and societal characteristics as a yardstick in the assessments of radiological impacts in the future. Accepting that knowledge of the future is limited, the effort spent needs to be adapted to the level of hazard/risk remaining and the progress that can be expected. It could be considered ethically questionable to spend a large part of the resources available to the current generation to study and prevent risks that remain highly hypothetical and are only expected in a distant future. As stated by one participant, many significant changes may occur that greatly modify the characteristics of society, man and the surface environment. It may well happen that we will appear to human beings of the far future much like apes appear to us today!

These latter comments lead us to examine two main issues from a scientific point of view.

Scientific viewpoint

The first issue relates to the credibility of earth and materials sciences and the possibility or the justification we may have for saying anything regarding a far future. The second issue relates to the existence of a time period after which the hazard associated with the disposal will have become sufficiently low to consider the protection objective as definitively achieved.

Obviously the difficulties in demonstrating safety increase whilst the credibility of our efforts to do so decrease as the assessment moves to a farther future. All participants agreed that this is something that needs to be frankly acknowledged. What this statement implies in practice is more debatable. For some participants, it is not scientifically acceptable in a safety case to consider beyond one million years. Any attempt to do so would serve to provoke negative reactions from the scientific community and would thus undermine the overall confidence. For other, science as a tool is very powerful but it must be accepted that on this timescale scientific information is rendered imprecise and of greatly diminishing value by the mounting uncertainties. Other participants considered that, even if it is true that science has limitations, natural systems and geology gives proofs of long-term stability. Provided that the site has been adequately selected, the surrounding geosphere is not expected to change significantly over very long periods. What will change, however, are its boundary conditions. Based on an understanding about geological processes, it is then considered proper to at least roughly describe what can be expected in the different time frames, including the farthest ones. Even when it is thought probable that the repository will be eroded "in the end", there is no objective reason to totally ignore that point. The questions then are "when could this happen?" and "what can be said in terms of acceptability?" In this context it was pointed out that leaving localised concentrations of very long lived radionuclides into the far future is an unavoidable aspect of the 'concentrate and contain' approach that underpins deep geological disposal.

As illustrated in Neil Chapman's presentation, radioactive decay and comparison between a geological repository and a natural ore body constitute a basis to answer these questions. Activity and radiotoxicity plots indeed display a cross-over between curves associated with waste emplaced in a geological repository and curves associated with an equivalent amount of natural uranium ore, but interpretation and significance of these cross-over points remains subject to debate. There was a feeling that such comparisons must be used with care. It was pointed out that different types of plots are being used that lead to different crossover times, thus potentially weakening the confidence in this type of argument. The crossover points of activity or toxicity curves have limited meaning from the point of view of risk and safety. Indeed the hazard and radiotoxicity indices that have been used for this purpose notably neglect the influences of radionuclide mobility, different chemical environments and the overall containment characteristics of the repository system or the natural ore. These are thus considered as incomplete or biased indicators. Even when the plots suggest that the repository in certain important aspects has become comparable to a natural system, this does not necessarily indicate a return to unconditionally safe conditions. However, even in the light of these reservations, participants generally agreed that radioactive decay and comparison with natural systems such as uranium ores offer a basis that should be explored for the regulatory approach.

The harmful properties of the waste will dramatically decrease with time (notably in the first hundreds or thousands of years) and, at least for spent fuel, the activity content of the repository will progressively evolve to a content similar to uranium ores on a timescale of a few 100 000 years. Whether it is reasonable and justifiable to conclude that risk will then be strictly comparable to the risk associated with a natural system and thus acceptable, was a question subject to contradictory opinions. Some participants suggested that the target time of some 100 000 years derived from the crossover points was sufficient to recommend the definition of different types of requirements depending on the time frame of concern. They hence advocated that because of the eventual similarity between a repository and a natural ore body, regulatory compliance should essentially be based on the assessment of the first period of some 100 000 years, and only little consideration (if at all) given to the period beyond. Other participants recommended the use of different weights for different time frames and agreed it was justifiable to put a lower weight to the assessment of an impact in the far future. They were however reluctant to endorse the view that while the 100 000-year period corresponded to a meaningful reference, nearly no consideration was justified after this period.

Public opinion and public acceptance

Complementary to the question of handling disposal evolution in the far future, the participants of Working Group A also addressed the question of the necessary importance to be given to short-term periods notably in order to better answer public expectations. The participants expressed a general feeling that safety assessments have often placed too little emphasis on the initial period of several hundreds to a few thousands of years. They noted that the feedback they have received from the public and other stakeholders indicates that this lack of emphasis does not match public expectations and concern. The participants recalled the more specific interest that usually exists for the timescale corresponding to the next 2 or 3 generations. Experience gained internationally however also suggests that this general observation has limitations. The public is obviously not a single and uniform group but rather divides into multiple components with different opinions and interests. Differences are observed between individuals, local communities, opponent groups, scientific communities, political decision-makers, and so on.

On several occasions, members of the public have shown a strong interest in events and situations that are expected to occur only in a distant future. In Sweden and Finland, the effects of glaciation on a repository have provoked a particular public interest even though not related to short-term evolution. It seems therefore inappropriate to draw definitive and simple conclusions about the interest and concern of the public. It is all the more difficult because feedback from the public is generally still weak in this field. In several countries, the public has not really had much opportunity either to learn about the premises for safety assessments or to comment on the guidance given in the regulations as regards timescales.

3.   Shared views

It is possible to draw from the discussion a number of points that constitute shared views of the participants. They are collected below.

Ethics provides general goals and, in accordance with these goals, we strive to protect future generations equally to the current generation. There are however practical difficulties in defining the meaning of "offering the same level of protection as today" with respect to the far future. What can be aimed at, however, is to leave future generations an environment that is protective to a degree acceptable to our own generation and that ultimately will remain within the range of variability of natural radiation that exists today.

Adopting the strategy to concentrate and contain already implies the acceptance of a deposit of long-lived radionuclides in a far future. After sufficient time, however, activity and toxicity of this deposit has returned to levels similar to those found in natural systems. For the most important types of waste, this occurs after a time period of several 100 000 years. This value is usually derived from cross-over points on activity or radiotoxicity plots. Because of the differences in materials, chemical environment and radionuclide mobility, the information of such plots and their meaning in terms of risk to a population have to be cautiously interpreted. However, the comparison with natural systems provides a useful reference.

Safety assessments and compliance judgements will build on system understanding and should recognise the unavoidable uncertainties regarding long-timescales. There is no real justification from scientific and ethical arguments to prescribe "hard" time cut-off in regulations. Instead, there is a preference for a gradual shift in the weights given to the different arguments that comprise the safety case when judging compliance on longer timescales.

When judging the suitability of a disposal system, regulators and implementers should also place emphasis on the different time frames in proportion to the corresponding hazard of the waste. For high-level waste and spent fuels, the initial period of a few 1 000 years is of particular importance.

4.   Points that would benefit from further consideration

In addition to the shared views above, the group also identified points that would benefit from a continued discussion. They thus proposed to give further consideration to the questions of:

- Regulatory approaches to the use of complementary indicators, such as environmental concentrations and availability of radionuclides, in safety assessments, in particular when discussing the far future;
- Conflicting information on the interests of different segments of the public – e.g. local populations – as regards, e.g., timescales, levels and types of information and level of protection.

The group also suggested that the IGSC produce a booklet on the protection requirements over different timescales and the treatment of these timescales in regulations and safety cases.

Group B

BARRIER AND SYSTEM PERFORMANCES WITHIN A SAFETY CASE: THEIR FUNCTIONING AND EVOLUTION WITH TIME

Chairperson: Allan Hedin (SKB, Sweden)
Rapporteur: Sylvie Voinis (NEA, France)

Members: Eric Fillion (ANDRA, France); Siegfried Keller (BGR, Germany)
Phillipe Lalieux (NIRAS-ONDAF, Belgium)
Lumir Nachmilner (RAWRA, Czech Republic); Vincent Nys (AVN, Belgium)
Javier Rodriguez (CSN, Spain); David Sevougian (YMP, USA)
Jürgen Wollrath (BFS, Germany)

1.   Main questions discussed by the working group

The following six questions formulated by the Programme Committee were used as the basis for the discussions in Working Group B.

- What is the role of each barrier as a function of time or in the different time frames? What is its contribution to the overall system performance or safety as a function of time?

- Which are the main uncertainties on the performance of barriers in the timescales? To what extent should we enhance the robustness of barriers because of the uncertainties of some component behaviour with time?

- What is the requested or required performance versus the expected realistic or conservative behaviour with time? How are these safety margins used as arguments in a safety case?

- What is the issue associated with the geosphere stability for different geological systems?

- How are barriers and system performances, as a function of time, evaluated (presented and communicated) in a safety case?

- What kind of measures are used for siting, designing and optimising robust barriers corresponding to situations that can vary with time? Are human actions considered to be relevant?

2.   Context

Introductory presentations were given by Hedin and Lalieux to stimulate the discussions. The presentations demonstrated how different geologic media, repository designs and waste forms may lead to partly different treatments of timescales in a safety case. The KBS 3 concept with spent fuel in copper canisters surrounded by bentonite and deposited in granitic rock is in many respects different from the ONDRAF concept with both vitrified waste and spent fuel in steel canisters deposited in Boom clay.

The risk to have the discussions focused in a specific concept was avoided by taking advantage of the fact that main geologic media such as clay, granite, salt and tuff were represented in the working group.

Other aspects of the way the WG organised its work are pointed out in order to place the results into context:

- The discussions focused mainly on the normal evolution scenario whereas other scenarios/events were more briefly looked at.

- The working group focused mainly on heat generating waste.

It was noted that when discussing the different time frames, it is important to bear in mind the evolution of the hazard of the waste, which decreases significantly over the time periods of interest.

2.1   Time frames

A common ground for the discussions of timescales was sought through a subdivision of timescales into common phases, not meaning sequential phases or different time frames functioning in parallel. Generally the time after waste deposition should be divided into four main phases:

- the operational phase between emplacement and repository closure, extending over several decades;

- the thermal phase (e.g. extending for 1 000 years to 10 000 yrs) during which heat generated by the waste increases the temperature of the host rock;

- the isolation phase (e.g. extending for 10 000 yrs to 10^6 yrs) during which releases from the disposal system are low, but have to be assessed; and

- the geologic phase (e.g. >10 000 yrs or >10^6 yrs) during which the isolating capacity of the engineered barriers is no longer ensured.

The operational phase was not treated further since this was outside the scope of the Workshop. Regarding High Level Waste (HLW) the thermal phase is one of the main transients that could determine a time phase. Also chemical, mechanical and hydraulic transients induced by the presence of the repository as well as the natural system transients such as the climate cycle transients, could be used to define time frames in the repository evolution. Events like human actions and earthquakes may also induce transients at any time. The figure below shows how ANDRA has segmented the repository system with time in major processes.

Figure 1. **From ANDRA's presentation**

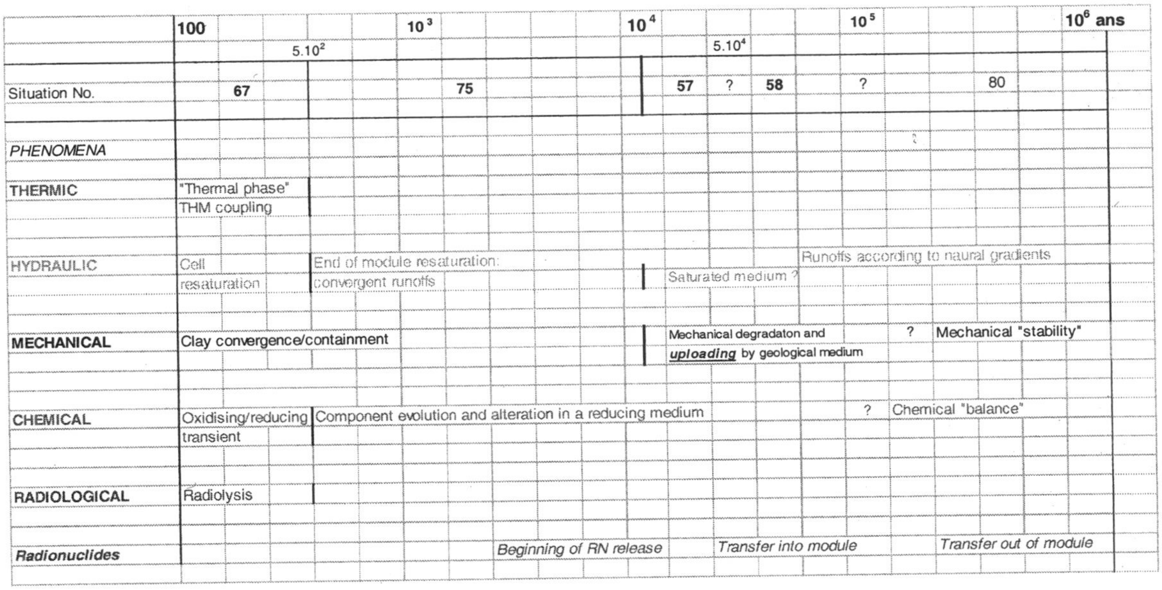

	100	5.10²	10³	10⁴	5.10⁴	10⁵	10⁶ ans
Situation No.	67		75	57 ? 58		?	80
PHENOMENA						?	
THERMIC	"Thermal phase" THM coupling						
HYDRAULIC	Cell resaturation	End of module resaturation: convergent runoffs		Saturated medium ?	Runoffs according to natural gradients		
MECHANICAL	Clay convergence/containment			Mechanical degradaton and *uploading* by geological medium		? Mechanical "stability"	
CHEMICAL	Oxidising/reducing transient	Component evolution and alteration in a reducing medium				? Chemical "balance"	
RADIOLOGICAL	Radiolysis						
Radionuclides				Beginning of RN release	Transfer into module	Transfer out of module	

Many common features that could be used to define time phases in a safety case exist between different repository programmes. These include similarities regarding waste forms, the fact that safety to various extents relies on an isolating barrier for which a life-time could be estimated and that all concepts discussed here build on deep disposal in geologic media which are expected to be stable over extremely long timescales. However, the detailed selection, duration and treatment of timescales in a safety case are strongly dependent on the particular disposal concept under consideration. Ultimately, it is the nature of the physical processes influencing repository evolution and the impact of these processes on repository safety that will to a large extent dictate the treatment of timescales in a safety case (cf. Figure 1).

3. The role of each barrier as a function of time

The long-term radiological safety is assured by the long term safety functions of the disposal system (i.e. the functions of the engineered barrier system and the host rock).

Several safety functions could exist in parallel whereas the safety demonstration may focus on only a selected number of functions, with the selection depending on the time phase, see Figure 2 below (de Preter & Lalieux: Part C). In the long term the main role is generally played by the geosphere, (including dilution in aquifers).

It was observed during the session that, the requirements of a component/function with time may be dependent upon the specific strategy, scenario class and/or time frame. Figure 3 shows an example of how different functions may be required from the same component depending on the safety strategy followed, in this case the different scenarios selected.

However, the handling of timescales often depends more on the strategy followed, the knowledge or confidence in knowledge available in the description of processes with time, rather than the type of components used. The nature versus characteristics of the common type of components (e.g. copper, iron, concrete, and clay) will depend on the concept (Hedin; de Preter & Lalieux; Bailey & Littleboy; Griffault & Fillion: see part C).

Figure 2. **The four phases of the normal evolution of the disposal system (high level waste) and the corresponding long-term safety functions. An example from ONDRAF[8]**

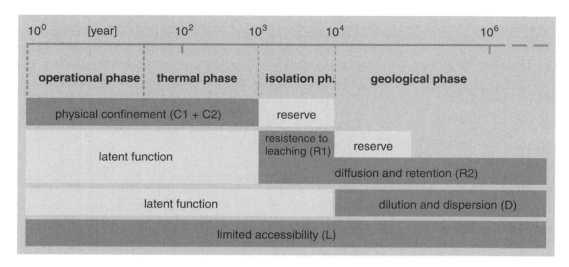

The biosphere is in general not regarded as a barrier since it cannot be optimised. However, the biosphere will generally play an important role in a safety case since it is an important factor in the quantitative assessment of the dose rate. It is important to assess how the favourable or unfavourable characteristics of the biosphere evolve on both short and long timescales, due mainly to climate and human behaviour changes. In particular dilution in the biosphere must be assessed in order to obtain a reasonable estimate of dose.

Figure 3:	Safety Functions in the KBS 3 concept assuming intact or defective canisters		
	Thermal phase 0 – 1000 years	**Isolation and Geologic phases; 0 – 10^6 years**	
		Intact canisters	*Defective canisters*
Fuel	As isolation phase	None	*Confine matrix nuclides*
Canister	As isolation phase	Isolate	*Limit release rate from canister interior*
Buffer	As isolation phase + Conduct heat	Protect canister (keep canister in position, prevent advective transport, exclude microbes)	*Retard* *Conduct gas H_2* *Filter fuel colloids*
Backfill	As isolation phase	Confine	*Limit advection in tunnels*
Geosphere	As isolation phase	Provide stable chemical mechanical	*Retard*
Biosphere	As isolation phase	None	*(Dilute)*

8. The long-term safety functions of the disposal system are "physical confinement" (C), "retardation & spread release" (R), "limited accessibility" (L); the long-term safety function of the environment of the disposal system is "dispersion & dilution" (D). The first two safety functions can each be subdivided in two sub-functions. For the "physical confinement" function, which aims to prevent any release of activity from the waste matrix the two sub-functions are "water tightness" (C1) and "slow water infiltration" (C2).

4. Main uncertainties in the performance of barriers

It was noted that this is a highly concept specific issue. However, a few observations of common nature were made:

- Regarding the EBS (Engineered Barrier System): many uncertainties concern the THMCR (Thermal, Hydro, Mechanical, Chemical, and Radiological) processes during the transient phases and therefore, their understanding might have implications for design. As an example, by using an overpack, treatment of uncertain, coupled processes influencing radionuclides transport during initial, transient phases can be avoided in a safety case.

- Geosphere uncertainties concern both the characteristics of today's geosphere and the long-term geosphere evolution including climate change, volcanism and seismic events.

- It is often stated that uncertainties increase with time, but this is seldom illustrated in safety reports.

- Engineered barriers may be altered during the operational phase and this is not always treated in the safety case. It is essential to both carry out adequate testing of barrier behaviour during this phase and to allow the test results to influence the design of the engineered barriers.

- Uncertainties might also come from slow process (chemical, physical) that could only have a significant impact after a long period of development.

5. Requested or required performance versus the expected or realistic or conservative behaviour with time?

The treatment of realism or conservatism/pessimism depends on the maturity of the repository program and the time frame considered and the final purpose of the evaluation. A conservative treatment may be acceptable, desirable or required when demonstrating compliance whereas a more realistic treatment is required for optimisation purposes and confidence building/model testing.

It was observed that the term "safety margin" is used both in a quantitative sense in comparing e.g. dose rates to a required regulatory level, and in a qualitative sense as a "reserve of safety" to account for safety functions which have not been taken fully into account in the safety case.

Quantitative treatment of the safety margin could not be reduced to a simple comparison of the doses to the regulatory criteria. It is an interesting tool not only to evaluate the discrepancy between the realism (best estimate) treatment and the conservative one but also for communication purposes.

6. Issues associated with the geosphere stability for different geological systems

It was noted that the concept of "geosphere stability" does not imply that steady state conditions are prevailing in the geosphere. It is important to assess resistance of the main safety functions (including the alteration of chemical, flow and transport properties) of the geosphere to man made or natural perturbation. In addition, natural observations could be used to enhance confidence in the geosphere stability and its predictability. The observation on the long-term diffusion in Opalinus clay presented by Nagra (see figure below) is an example of this:

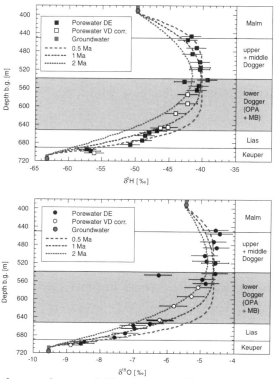

Also the crucial geosphere stability issues will be site and concept specific. Regarding for example the KBS 3 concept in Swedish granitic bedrock, intrusion of oxygenated water during a glaciation and impacts of earthquakes induced by rock stresses during de-glaciation are important questions in a safety case.

7. How are barriers and system performances as a function of time evaluated, presented and communicated?

Different evaluation methods exist to evaluate present and communicate the performance of a barrier as a function of time. The list below shows various types of arguments that could be used to evaluate performance:

- Thermodynamic arguments (stability of copper in Swedish deep ground waters)

- Kinetic arguments (corrosion rate of iron, fuel dissolution rate)

- Mass balance arguments (limited illitisation of bentonite, copper corrosion)

- Natural analogues (long term stability of bentonite)

- Palaeohydrogeology (long term stability of the geosphere)

- Long-term extrapolation of short term experiments/observations (corrosion processes, radioactive decay)

- Complex modelling (groundwater flow, radionuclide transport, earthquakes)

The validity of any of the types of arguments listed above depends on the extent to which they can be derived from a good mechanistic understanding of the physical process under consideration.

Safety and performance documents often focus on the impact of radionuclides releases by using the evolution of dose rate / or risk (or radionuclide flux) with time. However, in order to communicate performance of a particular barrier in accordance with its characteristics, radionuclide release is not always a practical performance indicator. As an example, regarding the isolation function required for the canister/overpack during the thermal phase, the evolution of the thickness of the canister /overpack with time might be a more convenient performance indicator.

The figure below also shows a different way to present the performance of each component with time rather than using the dose rate. The total inventory activity is given as well as the fraction of the activity present in each component in achieving the safety functions (Schneider, Zuidema & Smith: Part C)

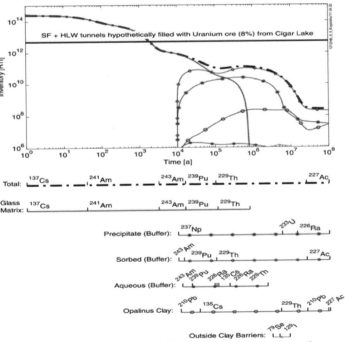

The advantages were recognised of illustrating barrier performance in different ways depending on the specific audience. For example the effects of processes (in which the time evolution of the disposal system is more easily checked and perceived) rather than the performance of barriers (in which the disposal system may look like oversimplified) may be more useful to people with a background in nuclear reactor safety.

8. Measures for siting, designing and optimising

Uncertainties concerning the long-term evolution of the repository system have influenced the siting and design of repositories in several ways. The following examples refer to the Swedish KBS 3 concept:

- A fundamental principle behind the design is the multi-barrier concept, through which the effects of a possible long-term deterioration of one barrier are mitigated by the presence of the other barriers.

- Another principle underlying the design of the repository is the choice of materials that are stable over long time periods. This applies to copper as the canister material and bentonite as the buffer material. Copper is thermodynamically stable under conditions expected in deep groundwaters in Sweden and bentonite is a naturally occurring material which was formed of the order of 10^8 years ago and which has since then mostly been exposed to conditions similar to those expected in deep groundwaters.

- The layout of the repository is chosen so that the distances to major fractures/faults, where future seismic events are expected to occur, are chosen so that the risk that these events will cause canister damage is kept low.

- A further design consideration is that the distance between neighbouring canisters is chosen so that the maximum temperature on the canister surface is kept well below 100°C. This is to avoid boiling and accompanying enrichment of salts on the surface which could in turn cause long-term corrosion effects that are difficult to analyse.

- Regarding siting, the repository will be located in a portion of the bedrock that does not contain ores of potential interest for future generations.

- A more general principle behind the deep repository is to isolate the nuclear waste from man and the surface environment, thereby minimising the effects of uncertain future societal changes.

9. Conclusions / main mutcomes

The term "multi-barrier" is used regularly in safety reports. However in doing practical work the term "multi-function" seems to be more adequate.

Regarding the terms "components" and "functions", a common list for all sites could be provided (e.g., canister, backfill, host rock, etc). In general similarities in functions and components exist over the various concepts or programmes, but their duration is often concept specific.

Terms like safety margins, robustness, barrier and safety function may have partly different meanings according to the various programmes. It is therefore important that clear definitions are given in the different contexts.

There are similarities of reasoning amongst various national programs to delineate time frames, repository barriers/functions, and key processes to fulfil safety and performance parameters. However, specific results are usually dependent on the stage of the repository program and are generally site and/or concept dependent.

Consensus exists on the need to split the discussion of post-closure safety into several time frames. However it is also noted that:

No "a priori" division in or duration of time frames is valid for all concepts and/or sites. Similarities or differences can be explained, e.g. during the thermal phase, by the similarities or differences in the various specific national programmes.

There are different rationales for division into time frames are e.g. predictability/ uncertainties (STUK "reasonably predictable future"), properties of waste (heat output, half-lives) and engineered barriers (durability, longevity), properties of site (climate cycles, transport times), and processes driving evolution (most of the preceding). In a national programme, it is important to clearly explain the rationale for the division chosen.

Based on safety strategy some choices (design safety function components) might also influence the choice of time frames. As an example, an intact canister (a design choice) could avoid the uncertain processes during the thermal phase and therefore if the knowledge of these processes is thereby improved, this could influence the design and usage of such a canister.

Group C

THE ROLE AND LIMITATIONS OF MODELLING IN ASSESSING POST-CLOSURE SAFETY AT DIFFERENT TIMES

Chairperson: Alan Hooper (Nirex Ltd UK)
Rapporteur: Gérald Ouzonian (ANDRA, France)

Members: Karl Göran Karlsson (SWD, Sweden); Jan Marivoet (CEN-SCK, Belgium)
Leonello Serva (ANPA, Italy); Paul Smith (SAM Ltd, UK)

1.   Introductory presentations

- Phenomenology dependent timescales: G. Ouzounian presented the PARS approach (Phenomenological Analysis of Repository Situations), through which ANDRA is structuring modelling of the repository. The principle of the method relies on a space and time segmentation of the repository evolution and a description of how it works;

- Long scale astronomical variations in our solar system: consequences for climate: K G. Karlsson has explained how from knowledge and understanding of the past it was possible to forecast the timing of major climate changes; it remains however difficult to determine the extent of each phase of those changes.

2.   Process of work

1. A quick review of our understanding and modelling capabilities of the stability of the main components of a repository was first performed, allowing for an initial idea on timescales to be considered;

2. A review of arguments for safety at different timescales, including the role of modelling, was then proposed;

3. Finally, an evaluation of the ability to handle "what-if" type questions has been conducted.

3. Review of understanding and modelling capabilities

Following the approach presented in the PARS, the group has segmented the repository system in major components, characterising each one with:

- the typical process which allows the definition of a characteristic timescale;

- a simple statement on our capabilities to model how it works;

- the limitations in understanding and/or modelling of the system behaviour.

3.1 Access tunnels

Typical process and related timescale: hydraulic transient for resaturation

Initiating processes: excavation of the tunnels and galleries, backfilling and sealing

Consequence: delay in achieving design properties

Statement: simple hydraulic process to be modelled

Limit: performance of seals on the long-term timescale

3.2 Waste emplacement tunnels: both short- and long-term processes could be identified

	On the short term	On the long term
Typical process and related timescale	THMC transients	Time specified functions : ▫ Physical: swelling, resaturation ▫ Chemical: sorption, buffering
Initiating processes	Excavation of the vaults, backfilling and sealing	▫ Interaction with water ▫ Temperature
Consequences	▫ Resaturation ▫ Heterogeneities, locally in space and time ▫ Delay in achieving design functions	▫ Changes in favourable properties of the buffer ▫ Changes in porosity
Statement	Validation of models depends on experiments. However, experimental evidence is incomplete (due to time and space scales and to the complexity of processes). Highly conservative design and/or improved R&D are to be developed interactively	Tunnels are designed to comply with the safety requirement according to the properties of both the waste form and the host formation
Limit	Behaviour of the system at scale 1 :1	Our capability to understand and to design; however the system can be bounded based on available knowledge

3.3 Container

Typical process and related timescale: failure time

Initiating processes: corrosion (alloys), ageing (concrete)

Consequence: loss of mechanical strength and mechanical failure, loss of confinement

Statement: as far as we understand the mechanisms and environmental conditions, we are able to define the failure time

Limit: design (thickness of the alloy, composition of concrete) and stability of environment

3.4 Waste forms

Typical process and related timescale: time during which radionuclides are still retained in the waste form

Initiating processes: contact with water

Consequence: corrosion and dissolution influenced by radiolysis

Statement: as far as we understand the mechanisms in the environmental conditions, we are able to define this time. If not sufficiently understood to have confidence in a realistic mechanism, can be bounded in a conservative way

Limit: environmental conditions

3.5 Geosphere

Typical process and related timescale: the timescale over which conditions will remain stable (driven by the same process) and no significant new features arise
Initiating processes: new features and deformations

Consequence: new conditions out of the boundaries of initial modelling

Statement: even for a well-chosen site, the possibility of new features and deformation must be considered over timescales in order of 10^5 years

Limit: quality of siting

3.6 Biosphere

The timescale for significant change is short. Initiating processes are many and largely unpredictable. The consequences for dose may be large but can be bounded. Nevertheless, the consequences for the boundary conditions of the disposal system are relatively small and reasonably understood.

The group is content with reference or stylised type biospheres.

3.7   *Surface environment*

Typical process and related timescale	Time to exposure of the waste to a direct influence from surface conditions	Time to onset of glaciation
Initiating processes	▫ Climate change (ice load) ▫ Tectonics ▫ Natural hazards (karst)	Climate change
Consequence	▫ Uplift-subsidence and erosion-sedimentation ▫ Evolution of pressure head and permeability	Evolution of pressure head and local faulting

Statement: as far as we can predict climate changes we can quantify consequences on the disposal

Limit: extrapolation from our understanding of the past (and of the processes).

4.   Summarised representation of the stability of the functions

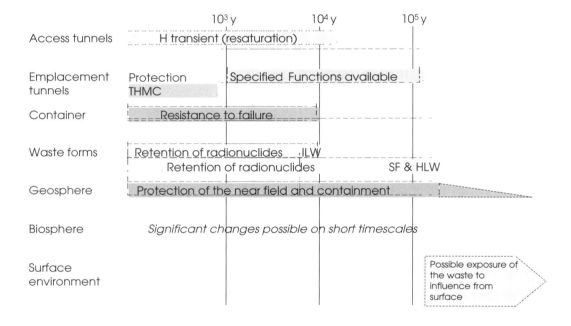

5.   Review of arguments for safety at different timescales

Q: What can we say about safety in the first thousand years?

- The most radiotoxic elements have decayed
- System is designed for confinement over more than 10^3 years

- Confinement is provided by well-known engineering materials, including arguments from natural and archaeological analogues

- Undisturbed host rock is well understood, but near field perturbations exist. Those perturbations may be irrelevant to radionuclide containment

- Arguments can be made using detailed models (e.g. container evolution) for a range of conditions

- Record keeping, monitoring and possibility of retrieval are considered within this time range

- Quality of site remains stable, with limited natural hazards

- Predictability of human actions is poorly predictable ; they are not relevant except for human intrusion

Q: What can we say about safety between 10^3 and 10^4 years?

- Stability of geosphere for a well chosen site can be demonstrated

- Quality of site limits natural hazards

- Most of the activity is confined within and near repository ; there is no migration over long distances

- Design of repository complements site characteristics ; design is developed to avoid complex and unfavourable conditions

- Understanding is reasonable to achieve support for release modelling (system is simpler than for the first thousand years

- Phenomena are decoupled from surface phenomena

Q: What can we say about safety between 10^4 and 10^5 years?

- Same as for 10^3-10^4 years:

- Stability of geosphere for a well chosen site can be demonstrated

- Quality of site limits natural hazards

- Most of the activity is confined within and near repository ; there is no migration over long distances

- Design of repository complements site characteristics; design is developed to avoid complex and unfavourable conditions

- Understanding is reasonable to achieve support for release and modelling

- Phenomena are still decoupled from surface phenomena

- Well supported models for radionuclide migration exist ; geosphere provides migration barrier and ensures slow migration

- Climate change : deep environment is largely unaffected for well sited repository (away from faults) ; support is given from understanding of geological and climate history

Q: What can we say about safety at timescales over 10^5 years?

- Decay has reduced the risk associated with waste. Radioactivity related to the waste has decreased at a level equivalent or lower than that of natural system

- Evolution of geosphere may be unpredictable or difficult to predict; it may be host-rock specific

- Consequences of glaciation and the following earthquakes are limited by the selection of the site, avoiding existing fault systems

- Timescale must be put in perspective with human history

- Calculations can still be made but with stylised representations

- Need to look at "worst case" (exposure at surface, to be compared to natural deposits, mill tailings)?

6. Handling "what-if" type questions

What-if type issues are mostly related to scenarios differing from normal evolution. An approach is required to address the question of "What" happens "if" an event occurs?

Given the approach which has been considered, the identification of different processes over the various timescales allows for a simple treatment of the what-if issues:

- Appropriate modelling capabilities are available; level of detail can be improved as far as useful

- Understanding of the state of the system at the time the event or perturbation occurs, allows modelling of the consequences of the perturbation

- System avoids or mitigates effects of many detrimental issues.

7. Conclusions from the working group

1) There is no universal solution to define timescales

2) The role of modelling is different at the different timescales, e.g. detailed corrosion modelling at early times. Selection of an appropriate approach to modelling is informed by consideration of stability. As a consequence, each time frame needs a specific approach and answer

3) Our understanding and modelling capability is such that each situation may have an appropriate answer from the basic knowledge of science and engineering.

4) For "what-if?" issues:

- Appropriate modelling capability is available

- Understanding the state of the system at the time "the perturbation" occurs allows modelling of its consequences

- The system can be designed and located to avoid or mitigate the effects of many detrimental issues.

Depending on the objective, modelling can be applied to describe the physical processes, according to the best available knowledge, or can be developed from this knowledge based approach through a decision based approach, typically to address:

- Scenarios, upscaling
- Stylised representations
- Uncertainties which are difficult to bound with R&D.

Group D

THE RELATIVE VALUE OF SAFETY AND PERFORMANCE INDICATORS AND QUALITATIVE ARGUMENTS IN DIFFERENT TIME FRAMES

Chairperson: Jesús Alonso (ENRESA, Spain)
Rapporteur: Klaus-Jürgen Röhlig (GRS Köln, Germany)

Members: Borislava Batandjieva (IAEA, Austria); Lise Griffault (ANDRA, France)
Alain Regent (CEA, France); Jürg Schneider (Nagra, Switzerland)
Richard Storck (GRS Braunschweig, Germany); Hiroyuki Umeki (NUMO, Japan)

1. Introduction

The Working Group adopted as a working method to organise the discussions around the questions provided by the Programme Committee and included in the programme of the workshop. The central topic in each question also provided the headings for the sections 3 to 6 in this report.

The exchange of opinions built on different source materials, namely:

- the presentations given in the plenary session of the workshop;

- the two introductory talks to the Working Group session;

- other relevant references, especially the relevant Technical Documents of the IAEA (e.g. TECDOC 767 "Safety Indicators in Different Time frames for the Safety Assessment of Underground Radioactive Waste Repositories", TECDOC draft of February 2002 "Safety Indicators, Complementary to Dose and Risk, for the Assessment of Radioactive Waste Disposal") and the on going C.E.C.'s Testing of Safety and Performance Indicators ("SPIN") project.

From the very beginning the working group was aware of two difficulties that could hinder the progress in dealing with the different issues under consideration:

- Different people may have various views on the definition of an indicator. Some may reserve this category for high level entities that provide an overall estimation on the functioning of a barrier, a sub-system or of the global system. For others, any entity at any level may be considered an indicator.

- There could be some expectations that the outcome of the Working Group would include a list of indicators with ready-to-use recommendations on their use.

55

The Working Group chose to rely on the general understanding of indicators as found in the references given above, with no need for further clarifications. On the other hand there was a general consensus on the need to avoid any prescriptive position that could raise a controversy. The potential for different indicators and applications appears large and diverse, as it is shown in the papers given in the Workshop: the relative merits of indicators are case specific in general. Nevertheless, the Working Group elaborated an array of indicators for the purpose of illustration, using elements included in the papers given during the Workshop.

Given the great and heterogeneous variety of what is usually called safety and/or performance indicators in PA's and other literature and the varying context in which these indicators are used, the Working Group decided to focus on a conceptual and qualitative discussion rather than to strive for specific recommendations regarding the use of additional indicators. Even though some discrepancy was identified with regard to the question which entities might be called safety or performance indicators and which do not belong to this category the group decided to avoid discussions on terminology. Instead, it was intended to cover all the types of entities being referred to as indicators in the references given above. Nevertheless, a need for clarification of questions concerning terminology and definitions was identified.

2.   Identification of performance and safety indicators

Indicators can be grouped into two broad categories:

1.   Performance and safety indicators calculated using the models and tools usually applied in performance assessments. They include a variety of entities that can be used to indicate safety and/or performance of different components or of the global repository system as for example:

   •   radiological dose or risk as the "traditional" endpoints of performance assessments;

   •   concentrations, fluxes and inventories of activity or radiotoxicity;

   •   the timing of events and processes relevant to safety and / or performance;

   •   most of these indicators are tested in the frame of the CEC project SPIN, (Storck & Becker: Part C).

2.   In-situ observations and measurements that can provide entities indicating favourable or less favourable features in the performance of a repository component, especially of the geological barrier. In addition, they can contribute to confidence building in model assumptions and provide evidences on barrier performance, which are independent from calculation results. Examples of such entities are:

   •   natural isotope profiles indicating a migration regime driven mainly by diffusion (Schneider, Zuidema & Smith: Part C);
   •   groundwater ages;
   •   paleohydrogeological information in general;
   •   the immobility of naturally occurring Uranium and Thorium in clay formations;
   •   natural analogues.

The availability of these indicators strongly depends on the specific site and repository concept under consideration. Usually no generic yardsticks and reference values are available for these entities.

For some indicators, however, it is possible to find reference values or "yardsticks", which define which are acceptable or "good" values of the indicators. In other cases the indicators cannot be compared to such a reference value; instead they may be evaluated at different points in time or in space to provide relative indications of performance, or finally they may convey by themselves a distinctive characterisation of performance (as transport driven by diffusion, in the example given above).

3. The (potential) use and usefulness of safety and performance indicators in different time frames and their contribution to confidence in the safety case

The Working Group members appreciated the enormous potential value of indicators additional to dose and risk, in particular in the framework of handling timescales, the very topic of the Workshop; this was seen as a view largely shared by the participants in it, given that virtually all the interventions during the plenary session gave considerable attention to this topic. As reflected by the Finnish regulations which acknowledge the different role of calculated dose in different time frames and require safety demonstration using calculated activity fluxes for the long-term (Ruokola: Part C), the relative value of certain indicators changes with time. Another example of such change with time is given in the paper by Bailey & Littleboy (provided in Part C of these proceedings) where data about the timing of container failure are used for early times to complement the information that dose is equal to zero and provide in this way information more meaningful about the performance of the system. In the long-term, when uncertainty increases remarkably, the relative value of qualitative indicators increases.

Indicators complementary to dose or risk can also support the communication of the findings of a safety case to different audiences. The comparison of impacts of a repository with that arising from naturally occurring nuclides might be helpful in this context. Messages concerning the confinement potential provided by the repository and the fact that it will not cause hazard to man and the environment can be supported using fluxes and concentrations, respectively (Umeki: Part C). Special emphasis should be put on the message that both entities will be equal to zero in the "early" phase in which some stakeholders seem to be especially interested in (Röhlig & Gay; Chapman: Part C).

However, confusion might be caused when the presentation of a safety case jumps from one indicator to another for different times. Therefore, a sound strategy concerning the choice and utilisation of indicators needs to be communicated. Presenters need to account for the fact that none of the indicators (including dose and risk!) and yardsticks to be compared to them are self-explaining.

The Working Group regarded the concept of reduction of uncertainties using indicators implied by one of the questions asked in the Workshop programme as somewhat questionable. What certain indicators do is to be free of certain uncertainties. It is in this way that indicators can increase confidence in both the performance of the system and the arguments of a safety case by providing multiple lines of evidence. Certain uncertainties which increase with time can be avoided by using specific indicators, e.g. the use of radionuclide or radiotoxicity concentrations in accessible water avoids the uncertainty coming from biosphere modelling and the use of fluxes avoids uncertainties about both the biosphere and dilution effects in the aquifer system (Storck & Becker: Part C).

As stated earlier, the Working Group acknowledged the increasing relative importance of qualitative arguments in the very long term when uncertainties concerning quantitative indicators increases remarkably. The comparison of the repository inventory to natural Uranium deposits is an example for such arguments (Ruokola: Part C).

The Working Group regarded the formulation of general recommendation concerning the value, role or use of specific complementary indicators in specific time frames to be inappropriate. Instead, they shared the view that these matters strongly depend on the context of the safety case under question and on the specific purpose an indicator is intended for. The derivation and utilisation of indicators requires an approach which has to be developed and tested. Views about a priori "good" indicators might well change during the development of a concept or a safety case and useful indicators are often not visible from the beginning. No single indicator is universal or perfect for any time frame. In any case, as important as the intrinsic merits of an indicator is the consistency of the argumentation where it is used.

4. Advantages, disadvantages and limitations of complementary indicators

The meaning of dose or risk calculated in performance assessment changes remarkably with time. When uncertainty increases and the reliability of dose calculation decreases, this meaning changes from expected performance to illustration. However, the Working Group regarded calculated dose or risk as a valuable and central information and therefore as a useful indicator for any time considered in a Safety Assessment. Therefore, the Group considered other indicators to be complementary instead of alternative to dose or risk.

Guidance concerning the advantages, disadvantages, and limitations of indicators complementary to dose or risk is available in several documents at international and national levels (Batandjieva, Hioki & Metcalf; Storck & Becker; Griffault & Fillion: Part C). However, these studies indicate a strong dependence on the context of the assessment under question. In any case, indicators need to be tested in this specific context.

5. Requirements on modelling tools and approaches resulting from the use of complementary indicators

In the case of indicators calculated in performance assessments, the SPIN project shows clearly that there are only very limited implications and requirements on modelling tools such as changes in input/output routines.

The use of measurements and observations for the building of confidence in either the performance of certain barriers or the validity of PA models usually requires certain ideas and approaches with regard to modelling. Again, these ideas and approaches are strongly dependent on the case and the system under consideration and no generic statements on this are achievable.

6. The definition of yardsticks or reference values for different indicators and implications on future work

Possible starting points for the definition of reference values are either considerations of acceptable hazard (as for dose and risk) or of negligible disturbance of nature. Examples for the latter are the work done by AECL and within the AMIRO project (Griffault & Fillion: Part C) where standard deviations of activity and radiotoxicity concentrations observed in nature serve as such yardsticks.

Problems concerning the use of entities observed in nature are the temporal and spatial scale at which the observations need to be made (on site, regional, world-wide?), the question whether or not natural conditions are "harmless" and the case of nuclides which are not or only at extremely low levels present in nature.

Within the SPIN project, the use of the radiotoxicity concentrations in drinking water as a yardstick for radiotoxicity concentrations calculated in performance assessments is an approach where both limited disturbance of nature and limitations on hazard (assuming that "natural" drinking water is harmless) serve as basis for the derivation of yardsticks (Storck & Becker: Part C). In the STUK regulations also both principles are applied by using a reference biosphere as well as natural conditions for the derivation of constraints of activity fluxes (Ruokola: Part C).

Biospheres which presently can be observed under different climatic conditions can also serve as natural analogues for the derivation of reference biospheres and therefore of yardsticks for activity fluxes or concentrations (Batandjieva, Hioki & Metcalf; Röhlig & Gay: Part C).

The Working Group agreed that the compilation work concerning natural concentrations and fluxes as it has been done under the co-ordination of IAEA (Batandjieva, Hioki & Metcalf; Röhlig & Gay: Part C) needs to be continued.

When the duration of processes or the time of occurrence of certain events or states serves as an indicator, yardsticks can be derived from natural processes, the timing of repository evolution or with regard to radioactive decay.

Certain indicators as e.g. fluxes through barriers or measurements and observations of nature are meaningful on their own and they may be useful even in the absence of yardsticks. Some members of the Working Group expressed doubts whether or not in the absence of yardsticks such entities can be called indicators.

7. The need for regulatory development

The Working Group agreed that regulations need to acknowledge and require the principle of multiple lines of reasoning and the use of indicators complementary to dose or risk and of qualitative information in a safety case. It was also agreed that regulations should not prescribe the use of specific complementary indicators but rather should try to give guidance on the provision of complementary information by the implementer. Only when regulations are not generic but instead concern a specific case, site or concept and after detailed studies have proven the interest and the applicability of eventual reference values (as in the case of STUK), they might well give specific requirements on the use of specific indicators.

8. Conclusion

The Working Group agreed that indicators complementary to dose or risk are of great importance for the provision of multiple lines of reasoning at different time frames and therefore for the building of confidence within a safety case and that regulations should acknowledge this fact. They are also of great value with regard to the understanding of the safety case by and the communication to different audiences.

The relative value of such indicators changes with time. For longer timescales qualitative information becomes more important. The meaning of calculated dose or risk is different for different timescales (ranging from expected performance to illustration) but dose or risk remains a valuable and central information for any time considered in a Safety Assessment.

Certain indicators (concentrations and fluxes) can provide information by avoiding certain uncertainties which increase remarkably with time (biosphere, dilution) but apart from that no generic opinion or recommendation can be derived since the value of specific indicators and the required degree of aggregation (over different nuclides or even of consequence and probability) strongly depends on the following:

- the motivation to build a safety case;

- the features of the Safety Concept;

- the specific properties of the site under consideration;

- the time frames of the assessment;

- the regulatory context;

- the audience the safety case is presented to (Yearsley & Sumerling: Part C); and

- the experience available to the assessment team.

PART C

COMPILATION OF WORKSHOP PAPERS

SOME QUESTIONS ON THE USE OF LONG TIMESCALES FOR RADIOACTIVE WASTE DISPOSAL SAFETY ASSESSMENTS

R.A. Yearsley
Environment Agency, United Kingdom
T.J. Sumerling
Safety Assessment Management Ltd., United Kingdom

1.   Background

In most environmental safety assessments, timescales are relatively short term, generally in terms of 10s of years and occasionally 100s of years, notably in climate change impact studies. These timescales are consistent with human experience and apart from climate change studies do not appear to significantly stretch scientific knowledge. Safety assessments for radioactive waste disposal appear to be unique in addressing the extremely long timescales, often measured in many 1 000s of years. This has become accepted practice for most radioactive waste disposal programmes. This paper examines the basis for consideration over such long timescales.

2.   Definition of the problem

A fundamental principle that underlies all national programmes is that radioactive waste must be managed safely. The preferred approach is the concentration and containment of radionuclides (IAEA 1995), although admitting the validity of dilution and dispersion for some waste types. The aim is to concentrate and maintain control of the hazard, initially by active and eventually by passive means, and thus minimise immediate and foreseeable future impacts to the public and to the environment. The strategy leads to a concentration of the potential hazard and inevitably to the possibility of impacts in the far future, after direct control over the waste has ceased. A key question is how long into the future should these impacts be assessed.

It is worth considering what we are trying to achieve through radioactive waste disposal programmes. Put simply, it is protection of humans and the environment. This view is supported internationally and is set out clearly in the International Atomic Energy Agency's Safety Fundamentals document "The Principles of Radioactive Waste Management" (IAEA 1995). The IAEA Safety Fundamentals document refers to the need for "an acceptable level of protection", which is further defined in different national regulations and regulatory guidance. For example, in the UK, the Environment Agencies' "Guidance on Requirements for Authorisation" (The "GRA") (EA 1997) includes the IAEA principle that:

> "Radioactive wastes shall be managed in such a way that predicted impacts on the health of future generations will not be greater than relevant levels of impact that are acceptable today."

This is linked to a requirement that:

"After control is withdrawn, the assessed radiological risk from the facility to a representative member of the potentially exposed group at greatest risk should be consistent with a risk target of 10^{-6} per year (i.e. 1 in a million per year)."

The timescale over which this risk criterion applies is not set in the GRA, which states that:

"No definite cut-off in time is prescribed either for the application of the risk target or the period over which the risk should be assessed. The timescales over which assessment results should be presented is a matter for the developer to consider and justify as adequate for the wastes and disposal facility concerned."

In the UK, it has generally been considered to be of the order of up to about 1 million years for a deep repository, (e.g. Sumerling (ed.) 1992). Timescales of this order seem to have become broadly accepted internationally although there are significant variations between different national regulations. The question then arises of whether the "acceptable level of protection" changes with increasing timescales. This question requires some consideration of the needs of different audiences and the motivations underlying the use of long timescales in safety assessment for radioactive waste disposal facilities.

3.   What are the potential audiences and their needs?

In recent years, it has become increasingly recognised that different audiences require different types of assurance in terms of safety arguments. Such arguments may be quantitative or qualitative and may involve alternative or complementary safety indicators in addition to estimates of dose and risk, which are the conventional regulatory safety criteria.

About 10 years ago in the UK, a quantitative performance assessment was seen as being at the heart of a safety case based on technical compliance with the regulatory requirements. In the intervening years, the importance of supporting arguments as a means of explaining the safety case to a wider range of stakeholders has grown. This is reflected to some extent in the regulatory guidance (EA 1997), which states:

"… a risk assessment provided by the developer is likely to be an important part of the post-closure safety case, although the relative importance of quantitative and qualitative arguments will change as uncertainties increase with the evolution of the disposal system over time."

"… a risk assessment or any other technical assessment is unlikely to be sufficient on its own to provide a satisfactory demonstration of safety. Although in general a developer will be expected to submit a quantitative assessment of risk, a risk assessment is likely to form only part of the overall safety case. Sufficient assurance of safety over the very long timescales which may need to be considered is likely to be achieved only through multiple and complementary lines of reasoning. … All the separate lines of reasoning which are mustered by the developer in support of safety will, in different ways and to different degrees, inform the regulatory decision."

In a wider international context, the need for additional supporting arguments alongside a quantitative performance assessment has been discussed, for example, in "Confidence in the long-term safety of deep geological repositories" (NEA 1999) and more recently in "Stakeholder confidence and radioactive waste disposal" (NEA 2000).

Part of the need for a broader range of supporting arguments seems to stem from recognition that different audiences are likely to interpret safety and any associated timescales differently. A range of audiences may be interested in the safety case for a repository; a few representative audiences and their possible interpretations of safety and timescale are considered here:

- **Political decision makers** have a significant role in establishing the legal framework for a repository development programme and in setting, or agreeing, the regulatory criteria by which safety will be judged. Political decision makers probably need to take the broadest view of the meaning of safety and timescales to accommodate a wide range of advice and concerns to achieve an appropriate degree of public acceptability for their decisions. In reaching such decisions, they need to take account of a wide range of inputs including advice from regulators, independent advisory groups and the broader academic community. In addition, they will need to take account of concerns raised by environmental pressure groups, issues of public acceptability at national and local levels and possibly any international implications that might arise in terms of protection beyond national borders. Thus, safety to a political decision maker might imply the relatively short term to meet the concerns of, say, a local population and at the same time, might extend into the medium future to be consistent with the principles of sustainable development.

- **Regulators** tend to look at the "technical" safety of a repository in terms of compliance with regulatory requirements. The regulatory criteria are likely to include dose or risk criteria as indicators of long-term safety and the regulations might prescribe a timescale over which such criteria apply or might, as in the UK, leave the timescale for safety assessment open to the repository developer to justify. In several countries there is a requirement to assess impacts up to a calculated maximum of dose or risk regardless of the time at which is occurs.

- **The scientific and technical community** is likely to view safety and timescales in terms of the credibility of science and engineering underlying the safety arguments. In the UK, over the past ten years, there has been a tendency to carry out assessments of deep sites up to the maximum time for which the models of evolution of a site could be supported from scientific evidence, i.e., taking account of climate change up to about one million years. In the USA the National Academy of Sciences recommended that calculations should be expected within the period of "geological stability" of the site. However, the timescales for which the evolution of different parts of the disposal system can be estimated based on science vary and there are also differences in predictability between sites. Thus scientific audiences are always likely to question whether existing scientific and technical knowledge can be extrapolated over the timescales used in a repository safety assessment. Uncertainties over the effects of low level radiation are also cited as pertinent to the discussion of safety, although this is not unique to assessment of solid radioactive waste disposal.

- **"The public"** is not homogeneous and will include a large number of different groupings and individuals. These may range from committed environmentalists operating at a national level, to local people affected by possible development of a repository, or to an individual with a particular concern about some aspect of radiation or radioactive waste. This heterogeneity in the public brings in a wide range of personal values to interpretation of safety and the timescale over which a repository should remain safe. These range from groups that will wish to explore quite sophisticated scientific arguments and exploit any scientific uncertainties concerning the long term, to groups who may take an interest only in the possibility of very direct impacts to themselves or their children over the next few tens or hundred years. Most importantly, "the public" can have a significant influence on political decision makers when judging the acceptability of any proposals for long-term management of radioactive waste.

The above comments indicate the broad range of expectations in terms of the definition of safety and timescales that a developer might need to satisfy in presenting a safety case for a repository.

4. What are the motivations underlying the use of long timescales?

A first consideration for the motivations underlying the use of long timescales is that of protection. This might be in terms of:

- **Protection of future generations** is one of the Principles set out in the IAEA Safety Fundamentals document (IAEA 1995), which states:

"Radioactive waste shall be managed in such a way that predicted impacts on health of future generations will not be greater than relevant levels of impact that are acceptable today".

This principle is carried forward into the UK regulatory guidance and probably underlies most regulatory approaches. No timescale is prescribed in the IAEA Safety Fundamentals document or in the UK guidance. From a human point of view, meeting this principle might imply protection over a timescale consistent with known human civilisations, i.e. a few thousand years, or protection over a timescale at least as long as Man, Homo Sapiens, has existed on the planet, i.e. about 100 000 years.

As generally interpreted, however, the focus of the IAEA principle is not on period of protection but rather on the level of protection and intergenerational equity. The demand that predicted impacts will not be greater than levels of impact that are acceptable today has been a very powerful one in practice. It has been taken to imply that attempts should be made to apply radiological protection systems developed to control present day and immediate future impacts to the assessments of impacts in the far future. Thus repository safety assessments have tended to focus on the very small doses and risks in the far future rather than the very high level of protection that are achieved over shorter timescales through containment and isolation.

- **Protection of the environment** is another Principle set out in the IAEA Safety Fundamentals document:

"Radioactive waste shall be managed in such a way as to provide an acceptable level of protection of the environment."

This might be interpreted as protection of the environment as a human habitat and resource and, in general, regulatory standards have been related to impacts on human health with the underlying assumption that protection of humans provided adequate protection of other species. This view is consistent with the ICRP general guidance (ICRP 1991) recently re-affirmed (ICRP 2000). There is, however, increasing interest in protection of species other than humans reflected in studies underway under the auspices of the IAEA (IAEA 1999), the ICRP and also under the European Union's 5th Framework Programme such the FASSET (Framework Assessment of Environmental Impact) Project (see www.fassett.org). The recommendations of these groups are, however, likely to focus primarily on protection from present day radiological hazards.

The UK regulatory guidance published in 1997 includes a requirement which states:

"It shall be shown to be unlikely that radionuclides released from the disposal facility would lead at any time to significant increases in the levels of radioactivity in the accessible environment."

This requirement

"… restricts the probability of significant local accumulation in the accessible environment of radionuclides released from a disposal facility and so contributes to the future protection of human populations and non-human species."

In practice, however, this has been interpreted as a secondary and less demanding requirement than that of meeting the radiological risk target.

Another possible interpretation is protection independent of a human presence, that is, a safe for all time argument. This argument has sometimes been advanced by environmental pressure groups and also sometimes in the political arena but, as far as the authors are aware, it has not been adopted as the basis for a regulatory approach.

- **Ethical concern** is one of the motivations underlying the principle of protection of future generations but another concern is equity of protection between generations. The ethical concerns underlying radioactive waste disposal have been discussed in some depth in an NEA publication (NEA 1995) and also in a publication sponsored by the Nordic countries (Nordic RP&NSA 1993).

These seem to provide the primary motivations that underlie the use of long timescales in radioactive waste safety assessments although a further, no less important, set of potential motivations also exist and these are largely based technical issues and to a lesser extent, on legal considerations. Such motivations include:

- **Applying a safety margin in face of uncertainty** – estimating performance over a very long timescale may offer some perceived benefits in terms of the safety margin available to account for future uncertainty in the evolution of the repository system. There is a difficulty in how such a safety margin might be assessed.

- **Radioactive decay and evolution of the hazard** – disposal of long-lived radionuclides might be considered as a driving force behind the use long timescales in order to account for radioactive decay and the consequent evolution in the hazard over time. Radioactive decay of long-lived radionuclides might imply timescales of the order of 100s of 1 000s of years. Consideration of the decay of, say, plutonium over ten half-lives would be consistent with this approach giving a timescale of around 250 000 years. Hedin (1997) has shown that the radiological toxicity of one tonne of Swedish spent fuel is on a par with the radiological toxicity of the natural uranium from which it was derived (i.e. 8 tU nat) after about 100 000 years, and this is proposed as a timescale for which a repository must function in the SR97 assessment.

- **Technical design tests** – these might be used to illustrate containment and isolation of the wastes, for example, through evaluation of the potential longevity of waste packages, backfills or seals under repository conditions, which might imply timescales of hundreds or a few thousand years. Alternatively, system tests might apply to consideration of the long-term stability of the host geological formation, which might imply a timescale of a 100 000 years or more.

- **Legal arguments** – estimation of the future longevity of the legal frameworks within which a repository is developed, operated and eventually closed is limited by the lack of knowledge about future societal stability. There are not many examples of laws dating back more than a few hundred years and this might imply that consideration legal arguments alone might imply timescales of any a few hundreds of years into the future. This would be consistent with the post-closure institutional control periods assumed in many safety assessments. In their regulations, the Swedish Radiation Protection Institute ask that doses be calculated to 1 000 years (SSI 1999), and this is based on the idea that legal arguments cannot be expected to prevail over longer timescales.

The above discussion only covers a number of possible motivations that might underlie use of long timescales in repository safety assessments although it seems reasonable to say that different motivations are likely to imply different timescales. If the needs of different audiences are also taken into account then it is unlikely that use of a single timescale is likely to be acceptable to al audiences. This would suggest the need for different emphases on different parts of the safety case to meet the needs of different audiences. This is happening in practice and can be seen in recent NEA reports. (NEA 1999, NEA 2000)

5.   Use of safety indicators at different timescales

Dose calculated on the basis of an expected evolution of a repository system is widely accepted as a basis for regulation. Dose may provide an indication of potential impact on human health for up to a few hundred years after repository closure – this is consistent with ICRP's view. (ICRP 2000). Over much longer periods, dose may be used as a quantity for testing against regulatory and design targets. It may be used for as long as regulatory or design requirement imply provided it is recognised that the estimated dose is only an indicator of safety and not a measure of safety.

In the UK, estimated radiological risk per year to a representative member of a potentially exposed group is used to provide a regulatory risk target. Risk is a theoretical quantity that can only be inferred or estimated but nevertheless provides a useful means for taking account of uncertainties over foreseeable evolutions and events that might occur over long timescales. As such, risk may be used as a quantity for testing against regulatory or design targets and thus provides a valuable indicator, but like dose not a measure, of long-term safety.

In addition to dose and risk, complementary safety indicators such as radionuclide fluxes and concentrations have received increasing attention in recent years, e.g. (EQ 2000). These might be used to indicate acceptable, or negligible, change to the environment, i.e. relative to natural geochemical fluxes. Such indicators have potential value in explaining safety arguments to non-technical audiences by providing comparisons of potential impact over long timescales with, say, current natural conditions.

Interpreting the results of safety assessments whether based on dose or risk raises a question of whether different weights should be applied to results at different times either in design or by the regulator. In terms of risk, does a risk of one in a million now mean the same at, say, 100 000 years? On face value, the risk should be the same but can this judgement be made on the basis of current scientific and technical knowledge? As timescales increase, the unquantifiable uncertainties are likely to increase and this might affect interpretation of the results of a safety assessment against a regulatory criterion, which is often expressed only as a single numerical value and not a distribution. A further question arises over whether there might be different levels of concern or perceived importance in results obtained at, say, 1 000 years and at 100 000 years. This question might relate to how different audiences interpret the meaning of safety at different timescales.

6.   How do long timescales affect credibility?

The credibility of a safety case for repository rests on gaining acceptance by the regulators, other experts and importantly, decision-makers and the public. The long timescales used in repository safety assessments go beyond normal imagination and this could affect credibility of a safety case with the public and with the scientific and technical community. Many members of the public would probably consider a 1 000 years as a very long timescale and times beyond that might not be seen as sensible or credible.

In the UK regulatory guidance, it is stated that:

"In general, assessments of the radiological impact of a facility should cover the timescale over which the models and data by which they are generated can be considered to have some validity. In the very long term, irreducible uncertainties about the geological, climatic and resulting geomorphological changes that may occur at a site provide a natural limit to the timescale over which it is sensible to attempt to make detailed calculations of disposal system performance. The timescale over which the Agency will expect to see detailed calculations of risk will therefore depend on the site and the facility and is a matter for the developer to justify. Simpler scoping calculations and qualitative information may be required to indicate the continuing safety of the facility at longer times."

This would appear to place some limitation on the extent of quantitative assessment required but the timescale is left open for the developer to justify on the basis of site characteristics or hazard posed by the waste. Importantly, the guidance recognises the possible need for simpler scoping calculations and supporting qualitative information as indicators of safety at longer timescales.

Consideration of the question of credibility leads to the not unsurprising conclusion that safety assessment results, their purpose and their interpretation need to be explained carefully in presenting a safety case in order to maintain credibility with a range of audiences.

The use of long timescales imposes a requirement for rigour and robustness on performance assessment and supporting scientific and technical programmes that is not generally applied to other environmental issues. This gives strength to the arguments presented in safety cases for radioactive waste disposal facilities to meet regulatory requirements, and public and political challenges. At the same time, attempting to address long timescales might be seen by some as over-stretching scientific credibility. There is probably no easy way of balancing these two opposing views other than perhaps through dialogue with the "public" and the wider scientific and technical community, that is, those outside radioactive waste disposal safety assessment. Such dialogue might lead to different emphases on the timescales adopted in repository safety assessments.

7.   Concluding remarks

A few interesting and potentially useful points arise from the discussion given above:

- The possibility of impacts arising in the far future is an inevitable consequence of the decision to concentrate and contain the waste. That a repository can be sited and designed such that significant impacts are estimated to only occur at very long times in the future is a positive feature and measure of success of the waste management solution.

- No single timescale is likely to be acceptable to all audiences or for all the motivations that underlie the use of long timescales. Therefore a safety case will need to use different arguments, lines of evidence and calculations at different timescales.

- The fact that no time cut-off is stated in a regulation need not be taken as a demand that impacts should be calculated over all times. Rather, common sense should be exercised and if results to long timescales are presented then the associated uncertainties and their interpretation should be explained.

- Safety indicators, which focus on the success of containment and isolation of the majority of radionuclides, can be presented in parallel to the more usual indicators, which focus on the small radionuclide releases and consequent doses. Timescales related to overall toxicity of the waste may be used to put these results into perspective.

- Dialogue with the wider scientific community and the public could inform the safety case and the timescales adopted. This might not be easy to achieve, however, since the public is a heterogeneous group that will bring a wide range of personal values to any dialogue process.

- Dialogue might lead to different emphases on the timescales adopted in repository safety assessments, possibly with less emphasis on extremely long timescales.

REFERENCES

[1] EA 1997: Environment Agency, Scottish Environment Protection Agency, and Department of the Environment for Northern Ireland. Radioactive Substances Act 1993 – Disposal Facilities on Land for Low and Intermediate Level Radioactive Wastes: Guidance on Requirements for Authorisation. Environment Agency, Bristol, January 1997.

[2] EQ 2000: Potential natural safety indicators and their application to radioactive waste disposal in the UK, EnvirosQuantiSci, Report number QSL-6297A-1, 2000 (Available as a UK Nirex Limited contractor report).

[3] Hedin A 1997: Spent Nuclear Fuel – How dangerous is it? SKB report TR 97-13, Stockholm.

[4] IAEA 1994: Safety Indicators in different time frames for the safety assessment of underground radioactive waste repositories, IAEA-TECDOC-767, Vienna.

[5] IAEA 1995: The Principles of Radioactive Waste Management, IAEA Safety Series No. 111-F, IAEA, Vienna.

[6] IAEA 1999: Protection of the Environment from the Effects of Ionizing Radiation: A Report for Discussion, IAEA-TECDOC-1091, Vienna.

[7] ICRP 1991: 1990 Recommendations of the ICRP, ICRP Publication no. 60, Pergamon Press, Oxford and New York.

[8] ICRP 2000: Radiation Protection Recommendations as Applied to the Disposal of Long lived Solid Radioactive Waste, ICRP Publication No. 81, Pergamon Press, Oxford and New York.

[9] NEA 1995: The Environmental and Ethical Basis of Geological Disposal, NEA/OECD, Paris.

[10] NEA 1999: Confidence in the Long term Safety of Deep Geological Repositories: Its Communication and Development, NEA OECD, Paris.

[11] NEA 2000: Stakeholder Confidence and Radioactive Waste Disposal, NEA/OECD, Paris.

[12] Nordic RP&NSA 1993: Disposal of high level radioactive waste – consideration of some basic criteria, Report by the Radiation Protection and Nuclear Safety Authorities in Denmark, Finland, Iceland, Norway and Sweden, Stockholm.

[13] SKI/SSI/HSK 1990: Regulatory guidance for radioactive waste disposal – an advisory document, SKI Technical Report 90:15, Stockholm.

[14] SSI 1999: Health, Environment and Nuclear Waste – SSI's Regulations and comments. SSI report 99:22, Stockholm.

[15] Sumerling TJ 1992: Dry Run 3: A trial assessment of underground disposal of radioactive wastes based on probabilistic risk analysis – Overview, UK Department of the Environment report DoE/RR/92.039, London 1992.

AN APPROACH TO HANDLING TIMESCALES IN POST-CLOSURE SAFETY ASSESSMENTS

Lucy Bailey and **Anna Littleboy**
United Kingdom Nirex Limited

Abstract

Previous Nirex post-closure assessments of deep geological disposal have been based on the use of probabilistic safety analysis covering many millions of years. However, Nirex has also published an assessment methodology in which the assessment timescale is divided into a number of discrete periods of time (time frames). Nirex is currently at the stage of planning the next update to its generic post-closure performance assessment and is considering the merits of using an assessment methodology based on time frames, in order to improve links with operational assessments and the provision of advice on the packaging of wastes, and to encourage stakeholder dialogue.

This paper has been prepared as part of Nirex's aim, wherever possible, to "preview", or seek input from others on, its ideas for new work to generate discussion and feedback. It describes an evolution of Nirex's published assessment methodology and outlines how it could be applied in an updated post-closure performance assessment of the Nirex generic phased disposal concept.

One approach to defining the time frames could be based on the evolution of the repository system, with time frames being distinguished by the occurrence of certain significant FEPs (for example the period of institutional control, expected time of container integrity or the period for which geosphere stability can be assumed). This approach provides a clear link to the system evolution, but needs to acknowledge the uncertainty associated with the timing of certain FEPs and hence with the definition of the time frames. An alternative approach would be to define time frames to reflect the periods of interest to stakeholders. This latter approach provides a mechanism for stakeholders to make early inputs to the assessment process, but may not always be easy to integrate into the development of a safety assessment seeking to respond to technically-defined regulatory requirements.

Nirex is seeking to integrate the above two approaches into an assessment methodology that incorporates stakeholder concerns into a technically robust representation of the repository system and its evolution. The Nirex FEP analysis approach to safety assessment involves the development of a "base scenario", that describes the expected or "normal" system evolution; and a range of variant scenarios that each consider alternative system evolutions that could affect safety. Nirex is considering defining a series of "nested" time frames, defined on the basis of the key characteristics of the repository evolution, rather than with reference to elapsed time. The base scenario would give a view of the periods for which these time frames are judged to be applicable, but as the time frames overlap, stakeholders would be afforded the opportunity to form their own views on this.

Adopting time frames as part of a safety assessment methodology has a number of implications. The assessment approach for each time frame may vary. For example, in later time frames more qualitative methods may be more appropriate. This has the advantage of providing an assessment framework within which qualitative arguments, alternative indicators of consequence and natural analogues could take on a higher profile.

1.   Introduction

Nirex's mission is to provide the UK with safe, environmentally sound and publicly acceptable options for the long-term management of radioactive materials. As one option, Nirex has developed a repository concept for the phased disposal concept of intermediate-level (ILW) and certain low-level radioactive wastes (LLW). This concept, illustrated schematically in Figure 1, is designed to meet the necessary safety standards and ensure long-term environmental protection. The concept is generic, in that it could be applied at any suitable site.

Figure 1. **Schematic illustration of the Nirex phased disposal concept**

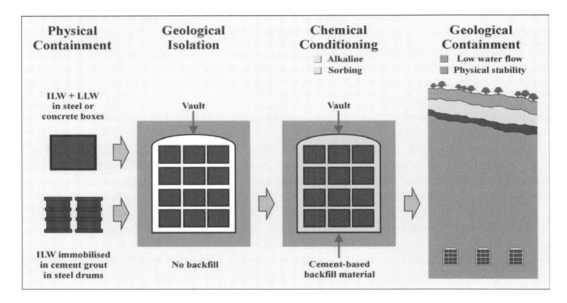

The Nirex phased disposal concept is a multi-barrier system, which envisages:

- immobilisation and packaging of wastes (physical containment);

- transport to a repository;

- emplacement in vaults excavated deep underground within a suitable geological environment (geological isolation);

- a period of monitoring during which the wastes would be retrievable;

- backfilling of the repository at a time determined by future generations (chemical conditioning); and

- sealing of the repository (geological containment).

In support of this concept, Nirex has published a series of Generic Documents, covering the design specification, operational, transport and post-closure safety assessments. These documents support the viability of phased deep geological disposal as a long-term waste management option and also support the packaging advice that Nirex gives to those organisations responsible for the UK's radioactive wastes. The advice is intended to ensure that raw wastes are packaged in a form that is suitable for protecting human health and the environment, both in the short-term, during storage, transport and handling operations, and over very long timescales following emplacement in a deep geological facility.

One of the suite of Generic Documents of particular relevance to this paper is the Nirex generic post-closure performance assessment (GPA) [1] which assesses the long-term safety of the disposal concept, following closure of the repository. Along with the other Generic Documents, this is a "live" document which will be updated periodically in line with new research and scientific, technical and social thinking. Nirex is currently planning the next update, which is due for publication in 2004. In line with its mission, Nirex recognises the need to ensure that this update integrates stakeholder concerns into a technically robust performance assessment.

Nirex has recently introduced the concept of "preview" into the development of new work programmes. Preview enables people outside Nirex to present their views on, and influence, Nirex work before it is carried out. This paper has been prepared as part of the preview process for the update of the GPA. It describes an evolution of Nirex's published assessment methodology [2] and outlines how it could be applied in an updated post-closure performance assessment of the Nirex generic disposal concept. It is anticipated that there may be further developments to the proposed approach, both in response to feedback from the preview process and during the implementation of the performance assessment.

2.   Assessments and stakeholder dialogue

Previous Nirex post-closure assessments have been based on the use of probabilistic safety analyses with a single set of time independent conceptual models developed to represent the repository system over many hundreds-of-thousands of years. The primary output of these assessments has been an expectation value of the radiological risk arising from a repository and its variation over time, for comparison with a quantitative regulatory risk target [3]. Recent analysis by Nirex [4,5] has suggested that this approach, whilst technically robust, may have hindered the discussion of post-closure assessment issues outside the expert community for a number of reasons:

- The quantitative regulatory risk target does not necessarily represent the concerns and values held by all stakeholders.

- Probabilistic assessment models have been used, in part, to attempt to account for a range of possible future system behaviours within a single model. Therefore the assessment models used can be very complex. This tends to focus discussion on the modelling process itself, rather than the underlying understanding.

- The quantitative presentation of results is rarely in a form aimed at engaging non-expert stakeholders. The long timescales involved are beyond the comprehension of most people.

Nirex's consultation with the public about their perceived hazards and concerns with the phased disposal concept [6] supports this analysis. Although some frequently asked questions relate to the post-closure period, most concern is expressed about the behaviour of the repository system during its operational period and in the time period immediately following repository closure. It has also become apparent that the previous assessment methodology (which, loosely speaking, considers how radioactivity could "leak out" so as to calculate risk) actually undermines the presentation and communication of the ideas underpinning the multi-barrier concept (within which the engineered and natural barriers both perform important containment functions).

Nirex experience suggests that there is public interest in discussing the safe management of radioactive waste [7]. However, the manner in which performance assessment is undertaken can act as a barrier to meaningful dialogue. To encourage discussion and dialogue about post-closure performance assessment, stakeholder concerns need to be addressed at the assessment methodology stage. Nevertheless, it is also true to say that a purely stakeholder-driven assessment would not necessarily generate a technically robust post-closure performance assessment and meet regulatory requirements. In order to communicate to both audiences using the same underlying performance assessment, the challenge therefore is to develop a methodology that can meet expert and regulatory requirements whilst demonstrably responding to wider stakeholder concerns. The balance between these two drivers will depend on the manner in which the performance assessment is to be used within a decision-making process.

3.　　　Evolving the Nirex assessment methodology

Nirex has developed a five-stage approach to performance assessment [2], as indicated in Figure 2. In this approach, a systematic analysis of all the features, events and processes (FEPs) relevant to the performance of the disposal concept leads to the identification of a base scenario and a number of variant scenarios that define potential evolutions of the repository. Each scenario is represented by a range of conceptual models, developed from knowledge of the FEPs operating in the scenario. For a detailed, time-dependent assessment, the time evolution of each scenario would be represented by a timeline, that is a sequence of key FEPs that divide the assessment timescale into a number of time frames. Within each time frame (also called a scenario interval), different conceptual models may be applicable.

This scenario development methodology has been published [8] and peer reviewed [9] but has not yet been fully implemented in a safety assessment as recent Nirex assessment work has been generic and has not explicitly considered the time evolution of the repository system. It is planned that the next update of the Nirex generic post-closure performance assessment should be based on the scenario development methodology, with particular emphasis on the explicit representation of different time frames. This introduces a series of practical matters that need consideration prior to implementation. These are considered further below.

Figure 2. **Nirex 5-stage assessment methodology**

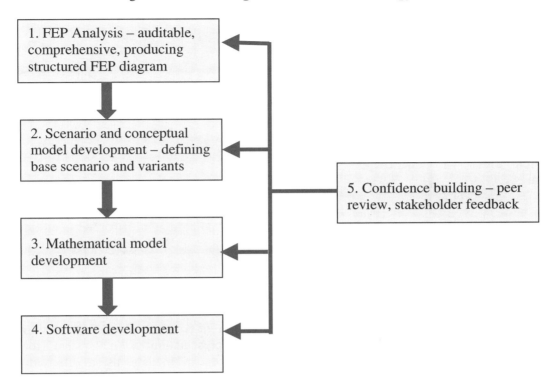

4. Time frame definition

One approach to defining the time frames could be based on the evolution of the repository system, with time frames being distinguished by the importance of certain significant FEPs (for example the period of institutional control, expected time of container integrity or the period for which the geosphere can be assumed to remain in more or less its current form). This approach provides a clear link to the changes in the repository system with time, but needs to acknowledge the uncertainty associated with the timing of particular FEPs becoming important and hence with the definition of the time frames. An alternative approach would be to define time frames to reflect the periods of interest to stakeholders. This latter approach provides a clear mechanism for stakeholders to make early inputs to the assessment process, but may not always be easy to integrate into the production of a safety assessment that has to respond to scientific and technically-defined regulatory requirements.

The challenge then is for an approach which simultaneously:

- accommodates the uncertainty in the timing of the impact of particular FEPs;
- presents a clear view of Nirex's understanding of the repository system evolution; and
- enables stakeholders to make inputs to the assessment process and explore a range of different assumptions regarding the time frames of interest to them.

It is believed the above requirements can be achieved if, rather than defining time frames sequentially, they are "nested". Nesting means that for modelling purposes each time frame is started at time zero, but the time frames extend for progressively longer periods, as indicated schematically in Figure 3.

Figure 3. **Illustration of sequential and nested time frames**

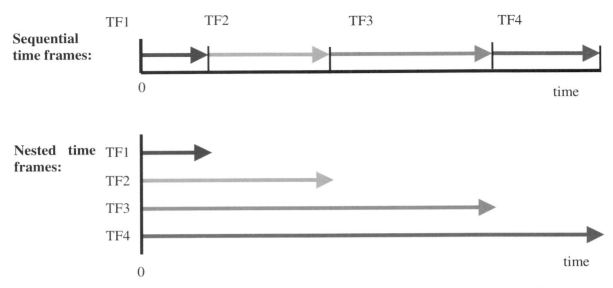

Dividing the assessment timescales into different time frames will provide flexibility to model each time frame in a different way. This will be true for both nested and sequential time frames. Each time frame can be considered as a different representation of the system (for example, encompassing different FEPs and hence different conceptual models, using different ranges of parameter values and even different assessment software). The aim is that by allowing this flexibility it may be possible to use tighter definitions of repository conditions, particularly for the earlier time frames, and hence produce a more meaningful and robust assessment.

Either approach to time frames would enable Nirex to weave some of the ideas it has been considering for many years into the fabric of the assessment (for example, alternative performance indicators, the use of natural analogues, confidence arguments, heterogeneity representation and new methods of risk communication), rather than just present them as add-ons. For example, natural analogues could be presented up front as arguments in support of the conceptual assumptions made for a particular time frame, and there would be the flexibility to look at different performance indicators in different time frames, rather than focusing solely on meeting a radiological risk target. This would shift the focus of the performance assessment more towards describing the factors that contribute to safety, rather than just calculating risk.

A key difference between a sequential and nested time frames approach is that with the former the decision on when to move from one representation of the system (time frame) to the next is made at the stage of defining the basis for the performance assessment, whereas with nested time frames sufficient information is presented to enable the assessment audience to make this decision if they so choose. This means that stakeholders can explore different possibilities by deciding when to make the transition between time frames. The sequential time frames approach would be less flexible to such "what if?" queries regarding the timing of particular FEPs as it would be necessary to define and assess a new time frame sequence (timeline) for each scenario.

However, whilst the nested time frames approach avoids the need to define specific start and stop times and consistent boundary conditions for each time frame, it will still be necessary to demonstrate consistency between the results from the various models at early times and to avoid the potential confusion of presenting multiple assessment results. These issues will be taken into account in the process of constructing the assessment.

5.   Constructing an assessment based on time frames

In view of the uncertainty regarding when particular FEPs might be of most importance, Nirex wishes to draw back from defining the time frames too tightly based on elapsed time. Instead, it is proposed that each time frame would be defined in terms of the aspects of the repository system that are contributing to its overall safety and the time frames will overlap and be open-ended.

However, it is considered that it will also be important to define a base scenario that gives a broadbrush description of a reasonable expectation of how the repository system may evolve. This base scenario would effectively be Nirex's expert recommendation of how the assessed time frames should be applied (as illustrated in Figure 4). It is suggested that the base scenario be defined on a cautious, but not overly pessimistic, basis. In addition to the base scenario, a number of variant scenarios would also be considered to deal with probabilistic events, such as human intrusion. This is consistent with Nirex's published methodology for scenario development [8], but avoids the need to formulate timelines with particular FEPs occurring at specified times. The base scenario would also be used to provide a reference basis for packaging advice.

Figure 4. **Construction of the base scenario from nested time frames**

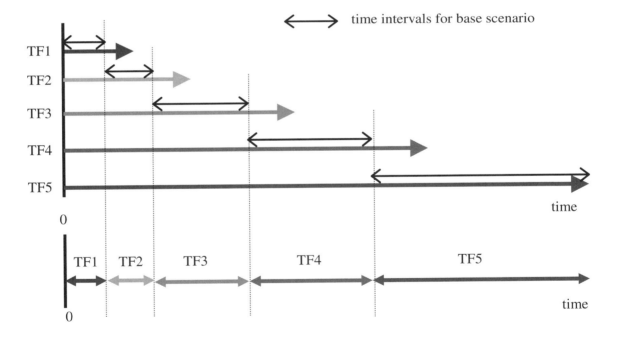

As an illustration, the following five time frames could be used to form the base scenario:

Time frame 1 –   waste containers as emplaced, repository backfilled and closed.

Time frame 2 –   physical and chemical barriers evolving, institutional control of repository site.

79

Time frame 3 – chemical barrier operating as designed, near-field reducing conditions established.

Time frame 4 – homogeneous near field, stable geological barrier.

Time frame 5 – system responding to external change.

In constructing the assessment, for each time frame consideration would be given to the following:

- The definition of the time frame, in terms of the importance of key FEPs relevant to changes in the repository system (the base scenario description).
- An indicative timescale (i.e. the period within the base scenario for which the time frame assumptions might reasonably be considered appropriate).
- Confidence arguments – natural analogues and other arguments, which support the scientific basis, adopted in the base scenario, i.e. those used to define the time frame.
- Relevant performance indicators.
- General modelling approaches, including:
  - appropriate model scales;
  - importance and relevance of different components of the repository system;
  - importance and relevance of different pathways.

- Potential variant scenarios for consideration. For example, human intrusion into the repository would need to be considered as a possibility from time frame 3 onwards, once there is no longer institutional control over the site.

The assessment would include a description of the base scenario. Assessment results would then be presented for each time frame (for example a chapter on each time frame). Each time frame chapter would include an explanation of the modelling approaches and their results, covering all the key impacts of the base scenario. Confidence arguments and comparisons with natural analogues would be interwoven in the discussion to explain how the repository system provides safety on the timescale in question. In particular, for the earliest time frame it would be emphasised that there is confidence the only releases from the waste packages would be gas. For this earliest time frame, the assessment results would focus on showing illustrations of the decay of levels of radioactivity within the waste packages.

Figure 5 provides an illustration of what is currently envisaged in terms of the assessment and presentation of each of the proposed five time frames in the planned update to the Nirex GPA.

The modelling approach proposed for time frame 5 is directly comparable with that used throughout Nirex's existing GPA [1], which for modelling purposes assumed that the near-field was homogeneous at all times, with no credit taken for containment by the packages or wasteforms. The results for time frame 5 would therefore provide a direct comparison with Nirex's previous assessment work, whilst the results from the earlier time frames will allow a more realistic evaluation of the roles of the barriers in containing the radionuclides at early times.

It is suggested that variant scenarios are similarly represented on a time frame by time frame basis, for each including a discussion of the likelihood of occurrence and the consequences if it did occur, in line with Nirex's published scenario development methodology [8]. In line with feedback on stakeholder concerns, a worst case scenario would be considered for each time frame. This would be presented in terms of all the things that would have to go wrong for the scenario to materialise, whilst explaining the mitigating features that would help to prevent each happening. This would ensure that these low probability scenarios are given appropriate conceptual consideration and are "visible" to stakeholders.

Figure 5. **Illustration of the presentation of an assessment based on five time frames**

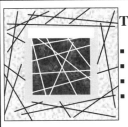

Time frame 1: Waste containers as emplaced

- Institutional control and monitoring (may include long-term storage period)
- Physical barrier intact, containment of radionuclides
- Repository starting to resaturate
- Releases limited to minor diffusive releases through vents and gaseous

Key Performance Indicators

- Decrease in radionuclide inventory
- Zero flux from near field
- Gaseous releases

Confidence arguments:

- Steel corrosion measurements
- Decay of short-lived radionuclides

Modelling approaches:

- Decay within packages
- Gas generation and transport

Time frame 2: Physical and chemical barriers evolving

- Repository fully saturated
- Physical barrier may start to break down but wasteform limits mobility
- Many radionuclides relatively insoluble, greatest release by diffusion
- Degradation of organics producing complexants
- Gas generation and migration

Key Performance Indicators

- Decrease in inventory
- Flux from near field
- Gaseous releases

Confidence arguments:

- Comparisons with corrosion of Roman nails
- Cement analogues

Modelling approaches:

- Package-scale model
- Near-field chemistry
- Gas modelling

Time frame 3: Chemical barrier

- Reducing conditions in near field fully established
- Corrosion causes failure of significant number of packages
- Advective-diffusive release of radionuclides, particularly those which are poorly sorbed
- Gas generation and migration

Key Performance Indicators

- Flux from near field
- Gaseous releases

Confidence arguments:

- Cement analogues
- Maqarin site – limited movement of radionuclides away from repository

Modelling approaches:

- Repository-scale near-field model
- Steady-state, regional-scale groundwater flow

Figure 5. **Illustration of the presentation of an assessment based on five time frames** (cont'd)

Time frame 4: Stable geological barrier

- Most waste packages have failed, offering little resistance to radionuclide migration, therefore the near field is treated as homogeneous
- Migration of radionuclides from near field through far field

Key Performance Indicators

- Fluxes out of near and far field
- Radiological risk
- Environmental effects
- Comparisons with natural fluxes

Confidence arguments:

- Maqarin site – limited migration
- Oklo – retardation
- Palaeohydrogeology – geosphere stability

Modelling approaches:

- Homogeneous near-field 'soup' model
- Groundwater transport models

Time frame 5: System responding to external change

- Homogeneous near field
- Migration of radionuclides from near field through far field
- Need to consider climate change and hydrogeological changes
- Releases to different climate states

Key Performance Indicators

- Radiological dose or risk
- Comparison with background radiation levels

Confidence arguments:

- Comparisons with natural radiation levels

Modelling approaches:

- Homogeneous near-field 'soup' model
- Reference geosphere
- Reference biospheres representing different climate states

6.   Regulatory implications

It is important that the proposed approach is consistent with regulatory requirements and capable of eventually being applied as part of a regulatory submission. Guidance on regulatory requirements in the UK is publicly available [3]. However, as with any regulatory guidance, it can be open to interpretation. Past performance assessments have been presented as focussed on determining consistency with the quantitative risk target laid down in the regulatory guidance. If this was all that was done, this could be seen as a very narrow interpretation of the guidance, which also sets out a number of qualitative principles such as the use of best science and the application of multiple lines of reasoning.

The assessment methodology proposed in this paper would remain consistent with past, quantitative approaches since it will deliver a risk calculation in time frames 4 and 5. However, it will also respond more explicitly to the broader considerations set out in the regulatory guidance by integrating evidence from analogue studies, underlying research and performance indicators other than risk into the overall approach.

7. Conclusions

It is believed that the methodology outlined in this paper for a repository performance assessment based on five nested time frames offers the following key advantages:

- It avoids the need to judge the exact timing of FEPs – time frames are defined on the basis of key characteristics of the repository evolution, rather than with reference to elapsed time.

- The approach is consistent with Nirex's published detailed FEP-based approach, but provides more flexibility in the context of a generic performance assessment.

- The use of nested time frames, with different modelling approaches in each, should enable a more accurate representation of the repository system for early times, whilst the approach proposed for the longest time frame provides a direct comparison with previous modelling approaches.

- Confidence arguments and the use of natural analogues are interwoven into the assessment, hopefully providing an overall assessment that will be more meaningful to stakeholders.

- The approach addresses stakeholder concerns by placing more emphasis on the early time frames and allows stakeholders to explore a range of "what if?" scenarios by presenting nested, open-ended time frames with different modelling approaches.

Nirex is investigating this revised assessment approach, in which calculations would be performed for a series of different conceptual models of the repository system. Each conceptual model would be most relevant over a particular time frame and the base scenario would describe Nirex's view on the most appropriate times to move between the conceptual models. However by nesting the time frames, stakeholders are afforded the opportunity to form their own views on the time periods over which they believe each conceptual model to be valid. It is therefore hoped that this time frames-based approach would not only provide the flexibility to take on board new trends in performance assessment, but will also be more meaningful to stakeholders.

REFERENCES

[1] L.E.F. Bailey, S. Norris, M.M. Askarieh and E.C. Atherton, Generic Post-closure Performance Assessment, Nirex Report N/031, 2001.

[2] L.E.F. Bailey and D.E. Billington, Overview of the FEP Analysis Approach to Model Development, Nirex Report S/98/009, 1998.

[3] Environment Agency, Scottish Environmental Protection Agency, Department of the Environment for Northern Ireland, Disposal Facilities on Land for Low and Intermediate Level Radioactive Wastes: Guidance on Requirements for Authorisation (Radioactive Substances Act 1993), HMSO, London, 1997.

[4] A.K. Littleboy, Performance Assessment as a Vehicle for Dialogue, Nirex Report N/037, 2001.

[5] A.K. Littleboy, Value Judgements, Performance Assessment and Dialogue: Report to the RISCOM 2 Project on Enhancing Transparency in Nuclear Waste Management. Nirex Report N/038, 2001.

[6] Identifying Public Concerns and Perceived Hazards for the Phased Disposal Concept, a report to Nirex by the Future Foundation, 2002.

[7] J. Hunt and P. Simmons, The Front of the Front End: Mapping Public Concerns about Radioactive Waste Management Issues, a report to Nirex by the Centre for the Study of Environmental Change, Lancaster University, 2001.

[8] D.E. Billington and L.E.F. Bailey, Development and Application of a Methodology for Identifying and Characterising Scenarios, Nirex Report S/98/013, 1998.

[9] OECD-NEA, Nirex Methodology for Scenario and Conceptual Model Development: An International Review, 1999.

CONSIDERATION OF TIMESCALES IN THE FINNISH SAFETY REGULATIONS FOR SPENT FUEL DISPOSAL

Esko Ruokola

Radiation and Nuclear Safety Authority, Finland

1. Introduction

Spent fuel disposal program in Finland has passed the decision-in-principle process that is crucial to the selection of the disposal site and to obtaining the political acceptance for the disposal plan. The regulator (STUK) participated in the process by reviewing implementer's (Posiva) safety case. The review was based on the general safety regulation [1] issued by the Government in 1999 and STUK's guide [2] of 2001 for the long-term safety specifying the general safety regulation. STUK's guide [3] for the operational safety of spent fuel disposal is being finalized. These regulations distinguish four time periods for which different safety criteria are defined; these are discussed below.

2. Operational period

The operational period involves encapsulation of spent fuel bundles, emplacement of the waste canisters into a repository at the depth of several hundreds of meters in crystalline bedrock, potential monitoring actions and the permanent closure of the repository. The operational period lasts typically some tens of years. In this time frame, spent fuel has very high potential radiotoxicity, but on the other hand, it is kept under strict human control. The safety criteria for the operational period, included in the Government decision [1], are as follows:

"The disposal facility and its operation shall be designed so that:

- as a consequence of undisturbed operation of the facility, discharges of radioactive substances to the environment remain insignificantly low

- the annual effective dose to the most exposed members of the public as a consequence of anticipated operational transients remains below 0.1 mSv and

- the annual effective dose to the most exposed members of the public as a consequence of postulated accidents remains below 1 mSv.

In the application of this Section, such radiation doses that arise from natural radioactive substances, released from the host rock or groundwater bodies of the disposal facility shall not be considered."

Demonstration of compliance with the safety criteria for the operational period shall to be based primarily on conservative deterministic analyses. However, for the justification the proposed safety approaches, adoption of PSA is also required, and its role increases in the later licensing phases. During the commissioning and operational period, extensive verification of compliance with the safety criteria by experiences and experiments is feasible.

3. Environmentally predictable future

The second time period, so-called environmentally predictable future is defined to extend up to several thousands of years. During this period, the climate type is expected to remain similar to that nowadays in Northern Europe. However, considerable but predictable environmental changes will occur at the disposal site due to the ongoing land uplift: a seabay will turn into a lake, then into wetland and might later on be used as farmland. The geosphere is expected to remain quite stable though slight, predictable changes will occur due to the land uplift and the heat generating waste.

In this time frame, the engineered barriers are required to provide almost complete containment of the disposed waste in order to minimize the impacts from waste induced disturbances and to facilitate retrievability of waste. Consequently, people might be exposed to the disposed radioactive substances only due to early failures of engineered barriers, such as fabrication defects or rock movements.

Despite the environmental changes, conservative estimates of human exposure can be done for this time period and accordingly the safety criteria are based on dose constraints. The Government's general safety regulation [1] includes the following the radiation protection criteria:

"In an assessment period that is adequately predictable with respect to assessments of human exposure but that shall be extended to at least several thousands of years:

- the annual effective dose to the most exposed members of the public shall remain below 0.1 mSv and

- the average annual effective doses to other members of the public shall remain insignificantly low.

Beyond the assessment period referred to above, the average quantities of radioactive substances over long time periods, released from the disposed waste and migrated to the environment, shall remain below the nuclide specific constraints defined by the Radiation and Nuclear Safety Authority. These constraints shall be defined so that:

- at their maximum, the radiation impacts arising from disposal can be comparable to those arising from natural radioactive substances and

- on a large scale, the radiation impacts remain insignificantly low.

The importance to long-term safety of unlikely disruptive events impairing long-term safety shall be assessed and, whenever practicable, the acceptability of the consequences and expectancies of radiation impacts caused by such events shall be evaluated in relation to the dose and release rate constraints specified in the Government Decision."

In the STUK guide [2], the radiation protection criteria are clarified as follows:

"The dose constraints apply to radiation exposure of members of the public as a consequence of expected evolution scenarios and which are reasonably predictable with regard to the changes in the environment. Humans are assumed to be exposed to radioactive substances released from the repository, transported to near-surface groundwater bodies and further to watercourses above ground. At least the following potential exposure pathways shall be considered:

- use of contaminated water as household water;
- use contaminated water for irrigation of plants and for watering animals;
- use of contaminated watercourses and relictions.

Changes in the environment to be considered in applying the dose constraints include at least those arising from land uplift. The climate type as well as the human habits, nutritional needs and metabolism can be assumed to be similar to the current ones.

The constraint for the most exposed individuals, effective dose of 0,1 mSv per year, applies to a self-sustaining family or small village community living in the vicinity of the disposal site, where the highest radiation exposure arises through the pathways discussed above. In the environs of the community, a small lake and a shallow water well is assumed to exist.

In addition, assessment of safety shall address the average effective annual doses to larger groups of people, who are living at a regional lake or at a coastal site and are exposed to the radioactive substances transported into these watercourses. The acceptability of these doses depend on the number of exposed people, but they shall not be more than one hundredth – one tenth of the constraint for the most exposed individuals.

The unlikely disruptive events impairing long-term safety, referred to above, shall include at least:

- boring a deep water well at the disposal site;
- core-drilling hitting a waste canister;
- a substantial rock movement occurring in the environs of the repository.

The importance to safety of any such incidental event shall be assessed and whenever practicable, the resulting annual radiation dose or activity release shall be calculated and multiplied by the estimated probability of its occurrence. The expectation value shall be below the radiation dose or activity release constraints given above. If, however, the resulting individual dose might imply deterministic radiation impacts (dose above 0,5 Sv), the order of magnitude estimate for its annual probability of occurrence shall be 10^{-6} at the most."

4. Era of extreme climate changes

Beyond about 10 000 years, great climatic changes, such as permafrost and glaciation, will occur. The range of potential environmental conditions will be very wide and assessments of potential human exposures arising during this time period would involve huge uncertainties. With conservative approach, the safety case would be based on extreme bio-scenarios and on overly pessimistic assumptions.

The climatic changes affect significantly also the conditions in the geosphere, but their ranges are estimable. In this time period, substantial degradation of the engineered barriers cannot be ruled out, though they were planned to withstand the stresses due to the climate-induced disturbances in bedrock. Radionuclide release and transport assessments in the repository and geosphere can be assessed with reasonable assurance and consequently, it is prudent to base the radiation protection criteria on constraints for release rates of radionuclides from geosphere to biosphere (geo-bio flux constraints).

The Government's general safety regulation [1] includes the following the radiation protection criteria for the era of extreme climate changes:

"The average quantities of radioactive substances over long time periods, released from the disposed waste and migrated to the environment, shall remain below the nuclide specific constraints defined by the Radiation and Nuclear Safety Authority. These constraints shall be defined so that:

- at their maximum, the radiation impacts arising from disposal can be comparable to those arising from natural radioactive substances; and

- on a large scale, the radiation impacts remain insignificantly low."

In the STUK guide [2], these criteria are clarified as follows:

"The nuclide specific constraints for the activity releases to the environment are as follows:

- 0,03 GBq/a for the long-lived, alpha emitting radium, thorium, protactinium, plutonium, americium and curium isotopes

- 0,1 GBq/a for the nuclides Se-79, I-129 and Np-237

- 0,3 GBq/a for the nuclides C-14, Cl-36 and Cs-135 and for the long-lived uranium isotopes

- 1 GBq/a for Nb-94 and Sn-126

- 3 GBq/a for the nuclide Tc-99

- 10 GBq/a for the nuclide Zr-93

- 30 GBq/a for the nuclide Ni-59

- 100 GBq/a for the nuclides Pd-107 and Sm-151.

These constraints apply to activity releases which arise from the expected evolution scenarios and which may enter the environment not until after several thousands of years. These activity releases can be averaged over 1 000 years at the most. The sum of the ratios between the nuclide specific activity releases and the respective constraints shall be less than one."

The selected approach means that the regulator has taken upon himself the burden of considering the biosphere impacts from the releases of disposed radionuclides, and the implementer need not to consider the bios-scenarios when preparing his safety case for the time beyond several thousands of years. The given constraints, which should still be regarded as tentative ones, were primarily derived on the basis of reference biosphere calculations. Besides that, some comparisons with fluxes of natural radionuclides in various scales were made in order to check the validity of the constraints and to have a more diverse standpoint on the issue.

5.   The farthest future

Beyond about 200 000 years, the activity in spent nuclear fuel becomes less than that in the natural uranium wherefrom the fuel was fabricated. In that time frame, the hazard posed by a spent fuel repository is comparable to that of a medium sized natural uranium deposits and the repository can be regarded as being part of the nature. No rigorous quantitative safety assessments are require for that time period but the judgement of safety can be based on more qualitative considerations, such as bounding analyses with simplified methods, comparisons with natural analogues and observations of the geological history of the site.

REFERENCES

[1]   General regulations for the safety of spent fuel disposal (1999) address disposal of spent fuel into bedrock, Government Decision 478/1999 (1999).

[2]   Long-term safety of disposal of spent nuclear fuel, STUK Guide YVL 8.4 (2001).

[3]   Operational safety of a disposal facility for spent nuclear fuel, STUK Guide YVL 8.5 (to be issued in 2002).

HANDLING OF TIMESCALES IN SAFETY ASSESSMENTS OF GEOLOGICAL DISPOSAL: AN IRSN-GRS STANDPOINT ON THE POSSIBLE ROLE OF REGULATORY GUIDANCE

Didier Gay
IRSN, France
Klaus-Jürgen Röhlig
GRS Köln, Germany

1.   Introduction

Due to the physical process of radioactive decay, the question of time is an inherent factor of any activity associated with nuclear materials. Hence time has naturally played a role in the definition of radioactive waste management strategies. This has eventually led to differences with other types of hazardous waste and notably to an earlier consideration of potential impacts in the long-term.

Time has therefore been early accounted for in the basic objectives of radioactive waste management. These objectives explicitly require the protection of human beings and of the environment **now and in the future**. In other words this implies the need to adequately manage the waste for "as long as it contains dangerous amounts of toxic substance", or, in the case of radioactive waste, "until sufficient decay has occurred".

For low and intermediate level waste (LILW), radioactive decay may be used as a powerful argument to determine the time-span needed for isolation. Therefore time has often been in the core of the strategy implemented for this type of waste. This has notably allowed for strategies such as the one adopted in France where a monitoring period of some hundred years for shallow LILW repositories is regarded as a key element for demonstrating safety.

In the case of high level waste (HLW) disposal, due to the long half-lives of some of the relevant radionuclides, the timescales of concern are comparable only to geological or even astronomical timescales and the potential advantage of radioactive decay is more questionable. Defining a strategy that remains sound and defensible all over the periods of concern is then a real challenge.

Traditionally, at the core of a safety case for radioactive waste disposal stands a Safety Assessment, which illustrates potential future evolutions of the repository system. Such illustrations become increasingly questionable with time due to several reasons:

1. **Increasing complexity**: the set of relevant possible evolutions (scenarios) grows larger with time since the number of relevant processes and phenomena increases and one possibly ends up with a nearly infinite variety of possible combinations which is hardly manageable.

2. **Limits to extrapolations**: even when processes and phenomena at stake are well identified and tools are available to assess their influence on the evolution of the system, extrapolations that these assessments require and validity of underlying assumptions often loose their sense with time. This applies especially for times when geological stability can no longer be assumed but there are examples for extrapolations which lose their validity even for earlier times (e.g. concerning hydrogeological or hydrochemical conditions influenced by climate changes).

3. **Factors non-amenable to prediction**: some other factors likely to influence future evolution of the system such as those related to human behaviour (human habits or societal processes) largely escape prediction even in the short term. In the lack of appropriate scientific background, attempts of prediction open large room for speculations. Two aspects are especially concerned: the definition of future human action scenarios (FHA) and notably possible human intrusion (HI), and the definition of environmental and societal assumptions necessary to calculate dose or risk to future population. For this type of factors uncertainties are clearly not reducible by science or implementation of research activities; it rather requires development of reasonable conventional approaches and search for consensus.

4. **Possible major disturbances in the far future**: many slow but continuous processes are at play and drive the evolution of the disposal system (e.g. predictable tectonic movements or uplift processes). Extrapolating these processes over long periods often lead to situations where the repository system is heavily damaged: in the far future, the worst seems to always remain thinkable. Even in a reasonably expectable evolution, the repository system might well end losing most of its isolation potential because of erosion of its rock cover occurring during major disturbances caused by tectonic events.

Two main types of difficulties finally arise: the first one relies in the wide range and variety of events and evolutions that need to be taken into account in the evaluation, the second one in the apparent inadequacy of the available knowledge with regard to the objective followed. They naturally lead to the following questions:

- What is the **adequate scope of a safety case**? How far in time must Safety Assessments go? What are the types of information to be produced for each different time frame? What are the use and the meaning of dose or risk calculations at these different times?

- How are difficulties coming from the **lack of a sound scientific basis** for the prediction of certain aspects to be faced? How can speculations concerning such aspects be avoided?

To overcome the difficulties encountered and help answering these questions, regulators are challenged to provide guidance, clarify the objectives or – if necessary – give more achievable ones. As technical organisations supporting the regulatory authorities in Germany and France, GRS Köln and IRSN have been both involved in this effort to provide guidance and clarifications. This paper presents their current standpoints on the subject.

2. Options for regulatory guidance

2.1 Options to define the scope of a Safety Assessment Case

Standing at the core of a safety case, Safety Assessment calculations are aimed at describing potential evolutions of the repository system and their consequences with regard to possible harmful effects on man and the environment. However, assessment calculations change meaning with time. Processes like changes in the aquifer system caused by glaciations (mostly expected to occur after some 10 000 years) or the loss of tectonic stability (or at least the loss of its predictability – for given sites after some millions of years), for instance, imply changes in the system under consideration which are hardly amenable to prediction using available models and data. The choice of the assessment time frame and, as an option, the definition of a time cut-off is therefore a central question for each assessment. The question of regulatory cut-offs is handled differently in different countries. Basically, two options are available.

In a first approach, calculations are closed using a "hard cut-off". One defines a point in time after which calculations do not need to be continued and no further consideration of longer time frames is required. This time cut-off may be justified by the argument that longer calculations would describe processes which are so uncertain that they nearly lose any meaning. Significant decrease of the radioactive content of the repository and accordingly significant decrease of its potential radiological hazard can be another type of justification.

The second approach can be referred to as "soft cut-off" approach. One acknowledges that calculated indicators like dose or risk may change their relevance beyond a certain period of time but contrarily to the first approach, this does not imply to close assessment or to completely disregard longer-term evolution. Justifications for this approach are multiple. First, it can be argued that calculated indicators might still provide useful information about the significant processes of concern even though calculations may become less relevant and do not predict them in a strict sense. Moreover, other types of information and arguments can complement them to give a reasonable idea of what could be the residual risk in the far future.

Soft cut-off is particularly justified as long as supporting elements are not considered sufficient to clearly identify a period after which no significant residual risk exists. Rather then prescribing a strict time scope of concern on relative subjective information, it leaves the possibility to discuss the acceptability of long-term evolution based on explicit information. Soft time cut-off may also help avoiding difficult situations where calculation results are abruptly stopped once the time cut-off is reached even though dose or risk curves are still rising and maximum values are not yet reached.

The view of IRSN and GRS Köln is that in the case of HLW disposal, time cut-offs can hardly be justified by considerations of radiotoxicity decreases or comparisons with uranium ore deposits. Such information provides valuable support for safety cases but finally, due to the very long half-lives of some of the radionuclides of concern and due to their high concentration in the waste forms, the waste may well remain hazardous for times beyond the usual cut-off times. Therefore, it is sensible for a safety case to provide the necessary information for a risk-informed decision for all the time frames of relevance. If it is clearly important not to overestimate the importance of late, highly hypothetical risks, it conversely seems unjustified to give no consideration at all to what could happen to the repository and what could be associated risks.

The choice of a "hard" or a "soft" cut-off itself finally relates to the type of information to be provided and to the particular place of dose calculations. Obviously the type of information and the importance of dose must depend on time. It is therefore necessary to recall that assessment calculations are not predictions but rather illustrations of the isolation potential of a repository. Even if they stand at the core of safety cases, they are not very meaningful on their own and have not the same relevance for different time frames of interest. It might well make sense to present a dose maximum occurring at 100 000 years when, supported by arguments about glaciation, a soft cut-off has been chosen at some 10 000 years. If, on the other hand, a 1 million years cut-off has been chosen since tectonic stability cannot be predicted for later times the presentation of a maximum occurring at 10^7 years can hardly be considered as useful information.

A sensible approach would be to carry out reference calculations by freezing present-day conditions and to use existing radiation protection objectives as yardsticks. Complementary to this information, the sensitivity of the results against climatic changes need to be evaluated by taking into account the influence of changing boundary conditions (e.g. on flow patterns) and of erosion effects. Calculations could be continued until dose / risk maxima are reached but the results have to be checked against the period for which tectonic stability can be predicted (i.e. the period during which the facility is not substantially damaged). If adequate, other arguments (qualitative or semi-quantitative) can be presented for later times. Safety cases need indeed to provide multiple lines of reasoning going far beyond such calculations.

In complement to the previous discussion, it is necessary to recall that such reasoning additional to the presentation of dose or risk curves may be very specific to the site, the safety concept, the type and quantity of waste considered and the purpose and the audience of a safety case.

2.2 Options for regulatory guidance where scientific basis is lacking or incomplete

2.2.1 Biosphere

The difficulty in modelling the biosphere lies in its extremely variable nature both as a function of space and time. As regards human society, the problem is even more acute. The pace of industrial development and changes in lifestyle proved to be continuous and significant since prehistory and is such that it can now be felt on an individual timescale. Therefore, a gap exists between the level of understanding we can hope to achieve regarding evolution in the geological environment over several tens to several hundreds of thousands of years, and the impossibility of predicting how the biosphere will evolve over the next few centuries. Handling timescales in this domain thus requires developing an approach to avoid pointless speculation.

Recently, the utilisation of performance and safety indicators alternative or complementary to dose or risk is increasingly being discussed. The use of geosphere fluxes or concentrations is recognised to have potential advantages to avoid difficult assumptions concerning biosphere treatment. However, a strategy completely relying on such indicators instead of dose or risk has also drawbacks with regard to both the definition of regulatory targets or limits and to questions of presentation and acceptance. If a Safety Assessment disaggregates the different radionuclides, both comparison with regulations and presentation become very complex. Also, the definition of yardsticks is not easy to achieve. If such yardsticks are defined based on conditions found in nature (e.g. natural concentrations or fluxes), three types of problems occur. First, it is not clear under what circumstances "natural" implies "non-hazardous". Second, it is not clear at which scale (site, region, country, world) such yardsticks should be derived. The choice of this scale could have unintended implications on the site selection policy. And third, it is not clear how to derive yardsticks for nuclides which are not or at very low concentrations present in nature. If, in contrast, yardsticks are derived using considerations on hazard, this might imply hidden biosphere considerations which one intended to avoid.

Where regulations rely on dose or risk and therefore imply biosphere considerations of some kind, the questions arise how complex such considerations need to be and, consequently, how credible they would be.

Implementation of "every" process known to drive biosphere evolution in the assessment would lead to a very complex task and would generate results that would very probably not solve the problem of credibility. It would be still full of speculations and very sensitive against parameters one cannot easily measure or derive. In addition, it would put huge emphasis on a part of the assessment which does not deal with the isolation potential of the repository system and is accordingly not in the core of safety concern.

If, in contrast, one restricts oneself to considerations about the drinking water well as it has been sometimes proposed, one would arrive with a very simple, understandable and not speculative model. However, studies show that the approach would mostly not come up with conservative or even representative results since significant pathways and radionuclides are not considered.

In the view of GRS Köln and IRSN, the most sensible approach is the derivation of reference biospheres. A reference biosphere is a collection of assumptions and hypotheses which are necessary for the derivation of a consistent basis to calculate radiation exposures. It should be as simple as possible but must not underestimate potentially significant exposition pathways. On this basis, a set of basic assumptions may be agreed on in order to avoid unnecessary speculation about future evolution. Possible hypotheses are that: technology, agricultural practices and characteristics of man remain constant; persistence of current climate or alternatively choice of climatic conditions among the range of climates expectable in the future; hypothetical critical group is a small farming communities living on local produce, crop production and animal husbandry/grazing. This type of simplified reference biospheres can provide a baseline information that can then be complemented and supported by scoping analyses which explore possible alternative biosphere notably those representative of climatic evolutions. Comparison studies that have been carried out recently both by GRS Köln and IRSN (the latter in the frame of CEC's SPA project [1]) give a good support to the adopted approach by suggesting that an international consensus on the subject is close and that this type of reference biosphere is a robust method to assess radiological impact.

2.2.2   Human intrusion

The simplest way to avoid speculations related to societal, economical and technological development and its implications on possible future human action in general and human intrusion in particular is to put them out of the scope of the assessment. This could be justified by arguing that human intrusion presents an unavoidable residual risk hardly manageable by design options but that provisions are taken to limit the likelihood of intrusion by choosing deep geological disposal as a management option and by choosing a site where such actions are unlikely. However, it is disputable that this argument constitutes a sufficient reason to definitely abandon further consideration to this type of situation. Such an approach can indeed be considered as non-defensible notably because it misses to address a situation that cannot be considered as unrealistic and that is additionally perceived by part of the public as one of the particular risks associated to geological disposal. An alternative approach is to acknowledge that human intrusion may happen in the future and to define a set of conventional or stylised scenarios that illustrate the robustness that can be expected from the geological repository. Such scenarios may be agreed on between applicants and regulators. Basic assumptions given in the NEA brochure about Future Human Actions [2] such as the exclusion of advertent intrusions and the consideration of present social habits and technology can easily be adopted as a starting point in order to avoid useless speculations. The conservation of knowledge about the repository in an initial period can be adopted as an additional assumption. The assessment can then be limited to inadvertent human intrusion scenarios occurring after a few 100 years. It is not yet clear to what extent consequences to the intruder need to be taken into consideration. However it appears preferable to give more particular care to the indirect impacts resulting from the intrusion and to check the ability of the disposal system to keep an adequate level of containment.

The choice of stylised human intrusion scenarios can be supported by compilation work about how different types of such scenarios have been treated in various Safety Assessments, the underlying assumptions and calculated consequences. Such work is being done in the framework of the present IRSN-GRS collaboration.

3. Present regulatory situation in France and in Germany

3.1 France

In France, main regulatory guidance relative to the disposal of radioactive waste in deep geological formations are currently found in the Basic Safety Rule n° III.2.f (BSR III.2.f) dated from 1991 [3]. Even if this text does not specifically aim to address the question of handling of timescales in safety assessments, it does contain relevant information.

A first valuable information is the absence of reference to time cut-off. The approach developed in the text is rather based on the consideration of a period of geosphere stability. The duration of this period is not a priori prescribed but rather need to be defined on a site-specific basis. The period is first referred to in the process of site selection. Site stability is thus considered as an essential criterion to examine site suitability. For a candidate site, it is explicitly required to demonstrate that the geological barrier will remain stable over a period of at least 10 000 years. The period of geosphere stability is further referred to in the definition of radiation protection objectives. As long as the geosphere is stable and for a "normal" evolution of the facility, the basic safety rule considers that assessment results may rely on objective data and can be backed by explicit studies of uncertainties. It is thus proposed to apply a dose limit of 0.25 mSv/year to judge the acceptability of radiological impacts. Beyond the period of stability, the activity of the waste decreases significantly. In the same time, the BSR III.2.f considers that uncertainty concerning the evolution of the repository becomes larger. The meaning and significance of dose calculations then change and the 0.25 mSv/year value shall be used as a reference value. They may then be supplemented by more qualitative assessments.

In addition to previous guidance, the Basic Safety Rules III.2.f proposes basis for the timing of the situations adopted for the purposes of the safety analyses. Three successive periods are distinguished. In an initial period of 500 years, it is considered that records of the disposal are likely to be kept and therefore human intrusion into the repository remains extremely unlikely. The value of 500 years shall be taken as the minimum lapse of time before intrusion might occur. This period is considered remarkable also in the way it corresponds to substantial decay of the activity of short and intermediate-lived radionuclides. An intermediate period, before 50 000 years, is then characterised by the absence of expected extensive glaciation. Lastly, during the subsequent period, after 50 000 years, it is consider necessary to make allowance in particular for extensive glaciation. This timing is further detailed in an appendix to the Basic Safety Rule where it is stated that 2 types of glaciations shall in fact be envisaged according to the Milankovitch theory. A Würm-type glaciation is first expected after 60 000 years, then a Riss-type glaciation after 160 000 years. For the Würm-type glaciation, French sites are expected to lie in a periglacial area and it is asked to evaluate the effect of the presence of permafrost or a fall in sea level on erosion and its possible consequences on groundwater flows. For a Riss-type glaciation, same type of evaluation shall be conducted but a conservative evaluation of the extension of ice is considered needed.

As developed previously a particular challenge raised by the handling of time in assessment is the necessity to avoid endless speculation notably by providing guidance relative to human intrusion and biosphere modelling. In this regard, the French Basic Safety Rule III.2.f indicates that the characteristics of man and society may be considered to be constant. It in particular means that sensitivity of man to radiation, the nature of food, contingency of life, general knowledge – including medical and technical fields – remain as they currently are. For biosphere, the Basic Safety Rule III.2.f explicitly states that it does not appear to be possible to predict the local changes in the environment over very long periods. It is however considered that major climatic changes are foreseeable at regional scale. It is thus recommended to apply the concept of reference biosphere that must be chosen as representative of the different states which might characterise the biosphere in the case of this expected evolution. Furthermore hypothetical critical groups are to be representative of individuals liable to receive the highest dose, including individuals living in at least partial self-sufficiency.

3.2   Germany

Presently, in Germany, the post-closure safety of a repository to be licensed needs to be evaluated on the basis of the "Safety criteria for the final disposal of radioactive wastes in a mine" issued in 1983. In these criteria it is stated that:

"after closure radionuclides which might reach the biosphere caused by transport processes which cannot completely be ruled out, must not lead to individual doses exceeding those mentioned in § 45 of the Radiation Protection Ordinance (Strahlenschutzverordnung StrSchV)."

The compliance with this criterion needs to be demonstrated using safety analyses for well-founded malfunction scenarios. The criteria do not make any statement concerning an assessment time frame. In 1989, the expert groups on reactor safety (Reaktorsicherheitskommission RSK) and radiation protection (Strahlenschutzkommission SSK) which are in charge of providing technical advice to the regulating authorities recommended to apply the dose criterion strictly to assessment time frames of less than 10 000 years. Since for times exceeding 10 000 years major changes in the system e.g. caused by glaciations are to be expected the commissions recommended to regard the dose criterion not as a limit but as a target for these times.

However, in the licensing procedure for the Konrad geological repository the licensing authority (the Ministry for the Environment of the Federal State of Lower Saxony NMU) did not adopt this view but instead required a compliance demonstration without defining any assessment time frame. In practise, calculations were carried out until the dose maximum was reached.

The regulations do not give any advice for the treatment of scenarios and the modelling of processes with a high potential for speculation, e.g. for those related to human activities and societal development. However, a technical guideline (Allgemeine Verwaltungsvorschrift AVV) is available that guides calculations of individual doses resulting from emissions of operating nuclear facilities. Even though conditions for the calculation of doses resulting in the long term from radionuclide releases from repositories are different from that and AVV is not strictly applicable, it still provides a frame for such calculations.

In the recent past, it was felt that the recent developments in science and technology related to Safety Assessments and safety cases cause a need for the development of new Safety Criteria. Therefore, on behalf of BMU a working group for the development of safety criteria (Arbeitskreis Sicherheitskriterien AKKrit) started to develop proposals for such criteria.

In parallel, as a result of the agreement between the power industry and the government, the question of siting for geological repositories has been revisited. On behalf of BMU, a working group develops new siting criteria and procedures (Arbeitskreis Endlagerstandorte AKEnd).

The development of both the safety and the siting criteria is still in progress. Several steps of development including discussions with stakeholders are foreseen. Therefore, the aspects presented hereafter are to be seen as preliminary.

The AKEnd defined it as a goal to search for geological settings where the part of the geology foreseen for enclosure of the radionuclides provides this enclosure for a period of geological stability of at least one million years. The view of the working group is that such geological settings can be found in Germany. However, for times exceeding several millions of years serious tectonic changes are to be expected and predictions going beyond these times are regarded to be completely speculative.

These views have natural implications on the development of safety criteria. The AKEnd considerations clearly address the undisturbed geological system and do neither account for disturbances caused by the repository nor for other disturbances of the expected evolution. The safety criteria in their present preliminary version state the validity of the protection objectives for an unlimited period of time but require an assessment period of one million years for both the expected evolution and the less likely scenarios. Beyond this time frame, no further consideration is required. With regard to the treatment of scenarios and the modelling of processes with a high potential for speculation, especially for those related to human activities and societal development, the criteria require the assessment of stylised situations and scenarios. In the case of human intrusion, the assessment should be limited to selected inadvertent human intrusion scenarios for which a conservation of knowledge about the repository for at least 500 years can be assumed. The dose calculations (biosphere assessment) should be undertaken based on reference biosphere models for the development of which a technical guideline is to be developed.

4.   Final remarks, summary and conclusions

The deep disposal of radioactive waste is a waste management option with some intrinsic qualities in comparison to other available alternatives. As stated in the RWMC collective opinion [4]

> "[…] this strategy would isolate the wastes from the biosphere for extremely long periods of time, ensure that residual radioactive substances reaching the biosphere after many thousands of years would be at concentrations insignificant compared for example with the natural background of radioactivity, and render the risk from inadvertent human intrusion acceptably small. Such a final disposal solution would be essentially passive and permanent, with no requirement for further intervention or institutional control by humans, although it may be assumed that siting records and

routine surveillance would in practice be maintained for many years if society evolves in a stable manner."

These qualities should be acknowledged in safety cases more clearly than it has been done in the past. Safety Assessments deal with the *remaining risk:* hazard has already been minimised by choosing the deep disposal option and disposing of the waste at an appropriate site and using an appropriate concept. These elements have to be borne in mind and can be used complementary to Safety Assessment results when defining an approach to handle timescale.

Lessons learnt during the public hearings and other discussions with stakeholders in connection with the licensing procedure for the Konrad repository in Germany as well as CEC's RISCOM-2 project show that the major concern of the public is often not on long term safety but rather on short-term aspects including transportation of waste, operation of the repository and protection of the living and the next one or two generations. Moreover, most people outside of the waste management community are neither familiar with the scope, objectives and methods of Safety Assessments nor have they much confidence in the ability of such assessments to demonstrate safety. Thus, the focus of most safety cases on the very long term does obviously not meet expectations outside the waste management community.

Regulatory guidance should acknowledge this by more adequately balancing the emphasis between the short-term including the early transient post-closure phase and the assessment of long-term safety. This is all the more important since stability in the early post-closure phase is an essential prerequisite for long-term robustness and safety. Therefore, repository concepts would obviously gain in safety if defined with the objective of generating as little as possible deviation from natural evolution and if based on barriers adequately understood as regards their influence on the overall robustness.

With regard to the latter, problems coming from increase of complexity with time and from limited abilities to extrapolate and to predict need to be acknowledged by regulators. This concerns both extrapolation and "prediction" based on natural sciences and areas where such "predictions" are fully speculative like in the case of societal evolution. It needs to be clarified that scenarios regarded in assessments are not predictions of the future but serve to illustrate the isolation potential of the system. Therefore, it is justified to regulate the handling of timescales in Safety Assessments in a way that both avoids calculations for times and with boundary conditions that are meaningless and overcomes problems linked with pointless speculations. The former can be done by choosing appropriate assessment time frames and giving guidance how to use and present calculation results respectively and the latter by defining stylised assumptions especially concerning the consequences of societal evolution based on current characteristics and taking into account evolution only for mechanisms that are well understood (e.g. climatic change).

REFERENCES

[1] Spent fuel disposal Performance Assessment (SPA project); EUR 19132 EN (2000); Baudoin P., Gay D., Certes C., Serres C., Alonso J., Lührmann L., Martens K-H., Dodd D., Marivoet J., Vieno T.

[2] Safety Assessment of Radioactive Waste Repositories. Future Human Actions at Disposal Sites. OECD/NEA; Paris, 1994.

[3]   French Basic Safety Rules — Rule N° III.2.f ; "Définition des objectifs à retenir dans les phases d'étude et de travaux pour le stockage définitif des déchets radioactifs en formation géologique profonde afin d'assurer la sûreté après la période d'exploitation du stockage"; 10 June 2002.

[4]   The Environmental and Ethical Basis of Geological Disposal. A Collective Opinion of the NEA Radioactive Waste Management Committee. OECD/NEA; Paris, 1995.

HANDLING OF TIMESCALES IN SAFETY ASSESSMENTS: THE SWISS PERSPECTIVE

Jürg W. Schneider and **Piet Zuidema**
Nagra, Switzerland
Paul A. Smith
Safety Assessment Management Ltd., UK

1. Introduction

This paper is intended as a contribution to the discussion of current approaches to the handling of timescales in safety assessments within national waste management programmes. It makes full use of on-going work within Nagra's Project *Entsorgungsnachweis*.[1] This project is based on a specific site; it considers a repository in the Opalinus Clay[2] of the Zürcher Weinland region in northern Switzerland. The repository is designed for the disposal of spent fuel (SF),[3] vitrified high-level waste (HLW)[4] from the reprocessing of spent fuel, and long-lived intermediate-level waste (ILW)[5] and is accessed via a ramp and a shaft. Emplacement of the wastes is in near-horizontal tunnels. Project *Entsorgungsnachweis* is divided into three sub-projects: (i) a synthesis of information from geological investigations of the Opalinus Clay, (ii) an engineering feasibility study and (iii) an assessment of long-term safety. Project *Entsorgungsnachweis* is a milestone in the programme for the disposal of SF, HLW and ILW and addresses an evaluation of the feasibility of the disposal of these wastes in Switzerland. It is also a major step on the way towards repository implementation. The corresponding reports will be submitted to the Swiss Government at the end of the year 2002.

2. Definition of key terms

There are two key project specific terms that need to be defined before discussing the strategy for handling timescales in the safety case for Project *Entsorgungsnachweis*. These are the *safety functions* and the *pillars of safety*; they are defined in the following two sections.

---

1. This German term translates into English as "demonstration of disposal feasibility".

2. Opalinus Clay is a shale (claystone) formation present in large areas of northern Switzerland.

3. The German term, used in Switzerland, is BE (abgebrannte Brennelemente).

4. Termed HAA (hochaktive Abfälle) in German.

5. Termed LMA (langlebige mittelaktive Abfälle) in German. This waste form is broadly similar to the waste category sometimes referred to as TRU – transuranic-containing waste – even though the transuranics may not be the most safety-relevant radionuclides in such waste.

2.1   Safety functions

The disposal system performs a number of functions relevant to long-term security and safety. These are termed safety functions; they include:

- *Isolation from the human environment:* The security of the waste, including fissile material, is ensured by placing it in a repository located deep underground, with all access routes backfilled and sealed, thus isolating it from the human environment and reducing the likelihood of any undesirable intrusion and misapplication of the materials. Furthermore, the absence of any currently recognised and economically viable natural resources and the lack of conflict with future infrastructure projects that can be conceived at present reduces the likelihood of inadvertent human intrusion.

- *Long-term confinement and radioactive decay within the disposal system:* Much of the activity initially present decays while the wastes are totally contained within the primary waste containers in the case of SF and HLW, for which the high integrity steel canisters are expected to remain unbreached for at least 10 000 years. Even after the canisters are breached, the stability of the SF and HLW waste forms in the expected environment, the slowness of groundwater flow and a range of geochemical immobilisation and retardation processes ensures that activity continues to be largely confined within the engineered barrier system (EBS) and the immediately surrounding rock, so that further radioactive decay takes place.

- *Attenuation of releases to the environment:* Although complete confinement cannot be provided over all relevant times for all radionuclides, release rates of radionuclides from the waste forms are low, particularly from the stable SF and HLW waste forms. Furthermore, a number of processes attenuate releases during transport towards the surface environment, and limit the concentrations of radionuclides in that environment. These include radioactive decay during slow transport through the barrier provided by the host rock and the spreading of released radionuclides in time and space by, for example, diffusion, hydrodynamic dispersion and dilution.

2.2   Pillars of safety

The pillars of safety are features of the disposal system that are key to providing the *safety functions:*

- *The deep underground location* of the repository, in a setting that is unlikely to attract human intrusion and is not prone to disruptive geological events;

- *the host rock* which has a low hydraulic conductivity, a fine, homogeneous pore structure and a self-healing capacity, thus providing a strong barrier to radionuclide transport and a suitable environment for the EBS;

- *a chemical environment* that provides (i) favourable conditions for the EBS and (ii) a range of geochemical immobilisation and retardation processes, and is stable due to a range of chemical buffering reactions;

- *SF and HLW waste forms* that have very low corrosion rates in the expected environment;

- *SF and HLW canisters* that are mechanically strong and corrosion resistant in the expected environment.

3. Strategy for handling of timescales in project *Entsorgungsnachweis*

The strategy for handling of timescales is derived from considering the key role of safety assessment, which is to provide arguments why the chosen repository system (the EBS and the host rock), at a given site, is adequate; i.e. sufficiently safe and sufficiently robust taking into account existing uncertainties. Next, it is helpful to bear in mind the limits of predictability of the various elements of a geological disposal system. This is illustrated in Figure 1, which indicates that due to the nature of the changes acting on these elements, the time over which predictions of their possible future evolutions are meaningful varies over orders of magnitudes as one moves from one element to the next. For a well-chosen site, a scientific database can, in principle, be obtained to support statements on the future evolution of the repository system for on the order of one million years. On the other hand, for radiological exposure modes, which depend, among other things, on individual human habits, predictions can, at most, be made for a few tens of years. As a consequence, and in contrast to the repository system, no thorough analyses are made for the biosphere, and the effects of uncertainties related to the biosphere are explored using stand-alone calculations yielding effective biosphere dose conversion factors.

Figure 1. Schematic illustration of the limits of predictability of a geological disposal system (NEA 1999)

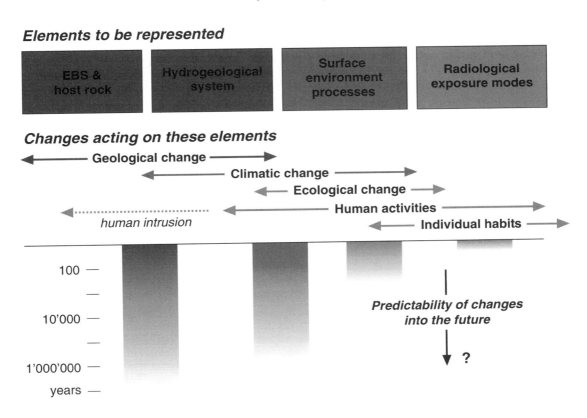

Another important piece of information relevant to the development of a strategy for handling timescales is the radiotoxicity of the wastes to be emplaced, given as a function of time, and as presented in Figure 2 for the three waste streams considered in Project *Entsorgungsnachweis*. The radiotoxicity is given in terms of a radiotoxicity index (RTI). The RTI of a given amount of radioactive material can be defined as the hypothetical dose resulting from ingestion of the material. Often, as is the case in this paper, the RTI is made dimensionless by dividing it by a reference dose; in our case this is 0.1 mSv (the annual dose corresponding to the Swiss regulatory guideline). In Figure 2, the RTI of the wastes is compared with that of the natural radionuclides contained in 1 km^3 of Opalinus Clay ("OPA")and with that of a volume of natural uranium ore corresponding to the volume of the SF/HLW emplacement tunnels. In the latter case, three uranium concentrations (uranium ore grades) are considered. These are 3%, which is the average uranium concentration of the small uranium ore body of La Creusa, Switzerland; 8%, which is a representative value for the Cigar Lake uranium deposit in Canada; and 55%, which is near the upper end of observed concentrations in uranium ore bodies. Figure 2 shows that after one million years, the radiotoxicity of even the most toxic waste, the spent fuel, has dropped to well below that of a volume of natural uranium ore sufficient to fill the SF/HLW emplacement tunnels. Figure 2 also clearly indicates that at times much beyond one million years, there are no additional benefits from isolating the waste from the human environment in terms of a further reduction of the radiotoxicity, unless one considers times up to 10^{10} years for which it becomes meaningless to make any kind of prediction of the evolution of the disposal system. As an absolute upper limit for such a time, one can consider the time corresponding to the age of our solar system (about 4.5×10^9 a). This is also roughly equal to the expected remaining lifetime of our sun (see, e.g., HOGAN 2000).

Figure 2. **Radiotoxicity index of spent fuel (SF), vitrified high-level waste (HLW) and long-lived intermediate-level waste (ILW) as a function of time, together with some reference levels corresponding to the SF/HLW tunnels hypothetically filled with natural uranium ore of different grades. The RTI of 1 km^3 of Opalinus Clay ("OPA") is also shown**

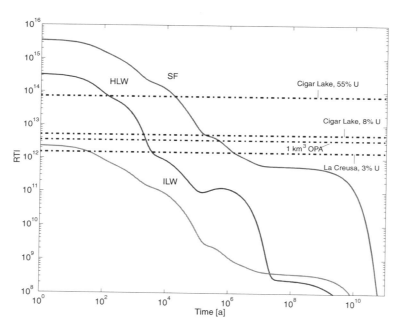

Thus one key element of the strategy for handling timescales in Project *Entsorgungsnachweis* is to place the main emphasis on times up to about one million years:

- There is a significant reduction of radiotoxicity of the wastes in this time frame and after one million years, the radiotoxicity is less than that of the emplacement tunnels filled with natural uranium ore (Figure 2);

- in the time frame of one million years, there is high confidence in key phenomena contributing to the safety functions (see Figure 3 for an example) and there is agreement in the scientific community that for a well-chosen site and a well-designed repository, an adequate scientific database can, in principle, be obtained that can be used as a basis for statements on the future evolution of the repository barrier system (EBS and host rock) up to about one million years (Figure 1);

- after about one million years, there is only little further reduction of radiotoxicity, unless one considers timescales that exceed the expected remaining lifetime of our sun (Figure 2); and

- statements become increasingly speculative.

Figure 3. **Isotope concentration profiles across the Opalinus Clay (OPA) and adjacent rock strata, comparing measured data (RÜBEL & SONNTAG 2000) obtained under various conditions (data points) and preliminary modelling results (GIMMI 2001) assuming diffusion only**

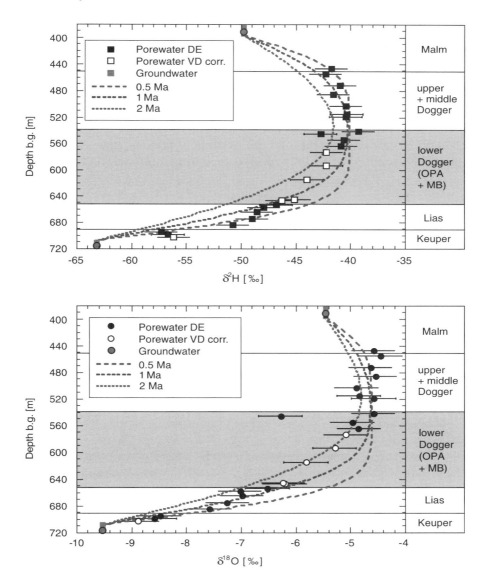

One key phenomenon contributing to the safety functions is the low groundwater flow in the Opalinus Clay host rock. Evidence for the high isolation capacity of the Opalinus Clay and its strong barrier function within the hydrogeological system of the sediment sequence comes from detailed analyses of pore water chemistry, and models of how this has evolved with time. Figure 3 shows measured concentration profiles across the Opalinus Clay and adjacent rock strata of the two naturally occurring isotopes, ^{18}O and ^{2}H, from porewater extracted from core samples from the Benken borehole. The isotopes are assumed to have been originally distributed approximately uniformly across the Opalinus Clay some few million years ago, when the current regional groundwater flow system was established. They have subsequently migrated outwards into the adjacent strata, modifying the uniform concentration profiles. The reason for the movement of these isotopes is changing water composition in the more permeable formations above and below, caused by flushing of the aquifers with younger waters. This caused a concentration gradient away from the centre of the host clay formation, allowing the isotopes to diffuse out into the Malm and Keuper aquifers. The best fit of the data was obtained by assuming that in the last one million years, outward diffusion of ^{18}O and ^{2}H was the only transport mechanism. Figure 3 provides compelling evidence for the dominant role of diffusion in both controlling porewater compositions in the host clay formations and, by analogy, in controlling the movement of any radionuclides released into those porewaters from a repository.

Figure 4 shows the key phenomena contributing to the safety functions, as a function of time. For the first 10 000 years (the design lifetime of the SF/HLW canister), radionuclides in SF/HLW are completely contained within the canisters. After canister failure, the group of phenomena termed "geochemical immobilisation/limited mobility/dispersion during transport" ensures that radionuclide concentrations in the surface environment and corresponding doses will be low. The horizontal dashed lines in Figure 4 indicate the redundancy of key phenomena: should, for example, a SF/HLW canister become breached at a time before the end of its design lifetime, the group of phenomena of geochemical immobilisation/limited mobility/dispersion during transport would start to operate at that (earlier) time.

Figure 4. **The approximate timescales over which various phenomena that contribute to the safety functions are expected to operate**

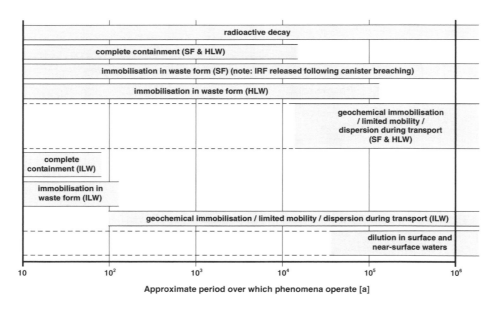

Figure 5. **Illustration of how most radionuclides are immobilised within the repository system, shown here for vitrified high-level waste. Note that at times before about 10^5 years, most radionuclides are immobilised in the glass matrix and at later times, most are sorbed in the buffer. Only a very small fraction is released from the host rock (labelled "outside clay barriers" and barely visible just before 10^7 years in the figure)**

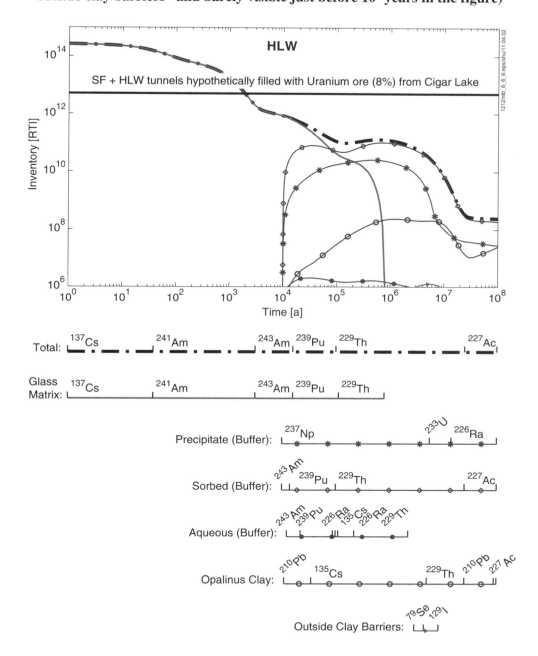

In the time frame from 10 000 years to one million years, most radionuclides are immobilised within the repository system (Figure 5). Some isotopes, however, such as I-129 from spent fuel, are not effectively immobilised, but their limited mobility and dispersion during transport ensures that releases from the host rock to the biosphere are so low that corresponding doses are well below the regulatory guideline.

4. Summary and conclusions

Nagra's strategy for handling timescales in the safety assessment supporting Project *Entsorgungsnachweis* is summarised in the following bullet points:

- A "reasonable" time for dose calculations is suggested to be on the order of one million years because for this time frame there is:

    - ... significant radioactive decay, with remaining radiotoxicity being comparable to that occurring in nature;

    - ... high confidence in the reliable operation of key (with respect to the safety functions) phenomena;

    - ... redundancy of key phenomena;

    - ... acceptance by the scientific community that an adequate scientific database can, in principle, be obtained to support statements on the future evolution of the repository system over this period of time.

- The assessment of system performance concentrates on the EBS and host rock, because:

    - ... it is these components of the system that ensure isolation and retention of radiotoxicity;

    - ... the geochemical phenomena that are key to the performance of the EBS are well-understood and stable over the period of interest;

    - ... the host rock – which is characterised by its low hydraulic conductivity and the absence of water-conducting features – is stable for the required time frame.

- The biosphere is modelled:

    - ... to illustrate the meaning of releases of a range of radionuclides from the repository system (Bq/a) by calculating an effective dose to an individual (mSv/a);

    - ... to assess the effects of uncertainties related to the biosphere by applying stand-alone calculations yielding effective biosphere dose conversion factors.

- For times greater than one million years:

    - ... statements become increasingly speculative;

    - ... the radiotoxicity of the wastes in the repository has become comparable to, or less than, that from natural materials;

    - ... there are no additional benefits from further detailed analyses much beyond one million years.

REFERENCES

[1]   Gimmi, T. (2001): Preliminary results.

[2]   NEA (1999): The role of the analysis of the biosphere and human behaviour in integrated performance assessments, OECD Nuclear Energy Agency PAAG document NEA/TWM/PAAG(99)5.

[3]   Hogan, C.J. (2000): Why the universe is just so, Rev. Mod. Phys. 72(4), 1149, The American Physical Society, October 2000.

[4]   Rübel, A. & sonntag, C. (2000): Sondierbohrung Benken: Profiles of pore water content, stable isotopes and dissolved noble gas content in the pore water of samples from argillaceous rocks. Nagra Internal Report, Nagra, Wettingen, Switzerland.

HANDLING OF TIMESCALES: APPLICATION OF SAFETY INDICATORS

Hiroyuki Umeki
Nuclear Waste Management Organization of Japan (Numo), Japan
Paul A. Smith
Safety Assessment Management Ltd., UK

1. Introduction

The H12 safety assessment (JNC, 2000) and subsequent safety studies of the possibility of deep geological disposal in Japan have considered the evolution of a hypothetical repository and its environment (the "disposal system") over the long timescale required for the calculated dose due to radionuclide releases to reach its maximum. Individual processes relevant to the evolution of the disposal system, however, occur over a wide range of timescales. At the one end of the range of timescales, changes within a well-chosen, stable geological environment are likely to occur on a timescale of some hundreds of thousands of years or more. At the other end of the range of timescales, resaturation is likely to occur on a timescale of a few hundred years. Even shorter timescales can apply to changes in conditions in the biosphere including the habits of potentially exposed groups. Assumptions must be made regarding these conditions, in order to evaluate dose, which is the primary safety indicator used in the H12 safety assessment (AEC, 1997). Many changes are only predictable to a limited degree and as a result uncertainties relevant to the safety assessment generally increase with time, as the range of possible paths of evolution becomes wider.

This paper discusses the treatment of uncertainty in the H12 safety assessment and, in particular, the different ways in which uncertainty in the evolution of the components of the disposal system, namely, the engineered barrier system (EBS), the geosphere and the biosphere, are treated. The use of safety indicators in the H12 safety assessment and, in particular, safety indicators that are complementary to the more standard indicator of dose, is also discussed.

2. Predictability of the components of the disposal system

In safety assessment, the objective is generally not to make precise predictions regarding the evolution of a proposed system, but rather to test whether the system is safe enough. A safety case can be made in spite of uncertainties, provided none of these uncertainties lead to the possibility of unacceptable radiological consequences. The approach taken in H12 was to consider separately uncertainties affecting the evolution of the EBS and the geosphere, and those affecting the biosphere. This separation is due to the very different role of the EBS and the geosphere compared to that of the biosphere in the safety case. The EBS and the geosphere are sited and designed to provide safety by ensuring that potentially hazardous materials are isolated from the surface environment and that any eventual releases of radionuclides to the surface environment are very small. The biosphere is relevant only in that it determines, to some degree, whether these very small releases comply with safety criteria, and particularly that for dose. The separation is further justified by the fact that, in many repository concepts, the evolution of the EBS and the geosphere is largely decoupled from events and processes occurring in the biosphere. Figure 1 shows the different components of the disposal system and illustrates the phenomena that may act on these components and the timescales over which predictions for the evolution of the components can reasonably be made. The biosphere (which is taken to include the habits or lifestyles of potentially exposed groups) and the upper parts of the geosphere are affected by a relatively large number of transient phenomena, limiting predictability. The deeper parts of the geosphere nearer to the repository and the engineered barrier system are predictable over much longer timescales. Although, for example, the next glaciation period is expected to occur around 10^4 to 10^5 years from now and will have a significant influence on the near surface environment and the hydrogeological system in particular, the repository can be sited and designed in such a manner that it will not be influenced by this or other disruptive natural phenomena for at least the next few hundred thousands years.

Figure 1. **The different components of the disposal system, the phenomena that may act of these components and the timescales over which predictions can reasonably be made**

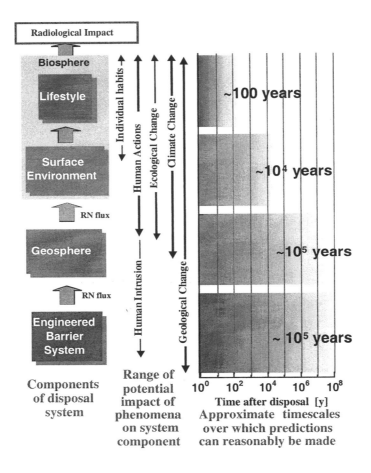

The treatment of uncertainty in the evolution of the biosphere in H12 is discussed in Section 3. The complementary safety indicators used in H12, and the yardsticks against which they were compared, are discussed in Section 4. The methodology for the treatment of uncertainty in the evolution of the disposal system has been extended since H12, and is summarised in Section 5.

3. Treatment of Uncertainty in the evolution of the biosphere

The evolution of the biosphere is realistically predictable over a timescale that is orders of magnitude shorter than those that apply to the other components of the disposal system. Human lifestyles are, realistically, predictable for no more than about a hundred of years at most. Furthermore, while it can be argued that some major features of the biosphere will probably remain comparable to those of the present day for up to around one thousand years, ecological changes in the biosphere due to the influence of human actions can be predicted for no more that some tens or hundreds of years. Any attempt at realistically predicting the evolution of the biosphere over the timescales addressed by safety assessments would amount to pure speculation. Nevertheless, if dose is to be used as an indicator of whether or not releases to the biosphere are acceptable from the point of view of safety, some assumptions regarding the evolution of the biosphere must inevitably be made.

The approach adopted in H12 was:

- to employ a stylised approach to defining conditions in the biosphere, using the concept of Reference Biospheres (e.g. BIOMASS, 1999); and

- to complement dose with other indicators of safety that are less sensitive to these conditions.

Reference Biospheres have been adopted internationally in various performance assessments in order to evaluate doses over long timescales. In H12, consistent with international practice, a Reference Biosphere was defined based on present-day environmental conditions and human lifestyles.

4. Use of complementary safety indicators in H12

4.1 Role of complementary safety indicators and yardsticks

Complementary safety indicators were used in H12:

- to give additional support to safety arguments based on dose calculations, by avoiding at least some of the assumptions inherent in these calculations;

- to provide safety arguments for a small number of "What-if" Scenarios for which the Reference Biosphere approach is inapplicable, namely releases due to uplift and erosion and to volcanic activity; and

- to illustrate containment, and the very high degree of radioactive decay that takes place within the repository and its immediate surroundings, instead of focussing solely on radionuclide releases.

The indicators themselves included radiotoxicities, radionuclide concentrations and radionuclide fluxes due to the repository. For each of these indicators, natural systems can provide yardsticks, allowing an evaluation to be made as to whether, for particular scenarios, the presence of the repository causes radiological perturbations that are significant in terms of what can be observed in nature.

4.2   Use of safety indicators that complement dose

Figure 2 illustrates how the output from geosphere transport modelling was used in H12 to provide not only radionuclide releases to the biosphere, as input to the biosphere model, but also radionuclide concentrations in groundwater and the concentrations of radionuclides sorbed on the mineral surfaces. These latter quantities were used as safety indicators to complement the doses calculated by the biosphere models. Yardsticks were provided by natural concentrations obtained from the analysis of groundwater and the analysis of core samples. These indicators may thus be regarded as independent of the assumptions of the Reference Biosphere.

Figure 2. **Assessment models, complementary safety indicators and the corresponding yardsticks**

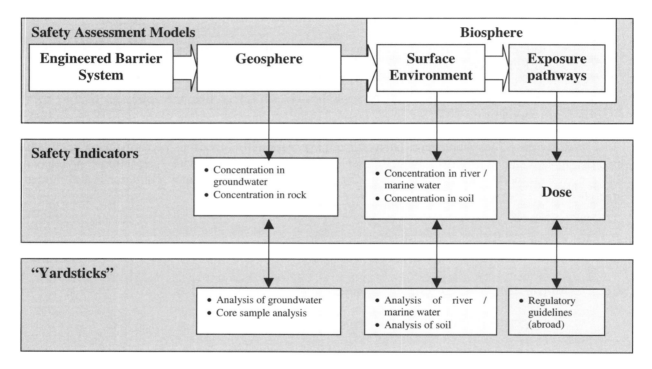

Similarly, the output of the surface environment model (a part of the biosphere model) was used to provide not only input to the exposure model that evaluates dose, but also radionuclide concentrations in river and / or marine water and in soil. These quantities were also used as safety indicators, with yardsticks provided by natural concentrations obtained from the analysis of river and marine water and soil. These indicators, although not entirely independent of the assumptions of the Reference Biosphere, are at least independent of assumptions regarding exposure pathways.

As a specific example, Figure 3 shows, as functions of time, the concentrations of ^{238}U and its daughters in river water resulting from radionuclide releases from the repository, calculated using H12 Reference Case releases to the biosphere and assuming the same river flow rate as in the H12 Reference Biosphere. The yardsticks here are the concentrations of the same radionuclides naturally present in river water, and also the maximum permissible uranium concentration for drinking water, according to the Drinking Water Quality Guidelines of the World Health Organisation (WHO, 1998). The figure shows that natural concentrations are not expected to be significantly perturbed by repository releases.

Figure 3. **The concentrations as functions of time of ^{238}U and its daughters in river water resulting from radionuclide releases from the repository, calculated using H12 Reference Case releases to the biosphere and assuming the same river flow rate as in the H12 Reference Biosphere**

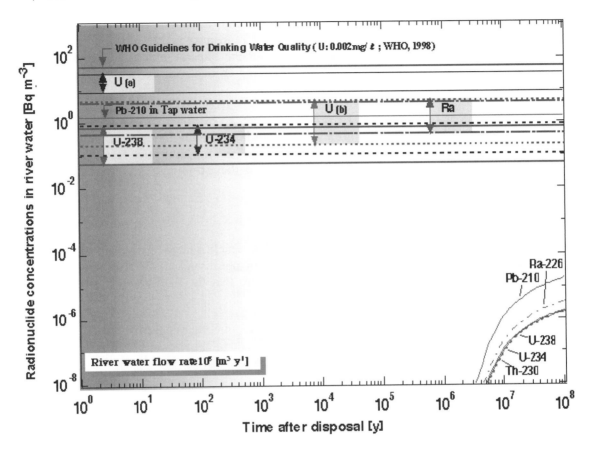

Reference: U(a) (Miyake *et al.*, 1964), U(b) (Tumura & Yamasaki, 1992), Ra (Miyake *et al.*, 1964), U-238 (Kametani *et al*, 1991), U-234 (Kametani *et al*, 1991), Pb-210 (Kametani & Tomura, 1976).

Such direct comparisons can, of course, only be made for radionuclides that are naturally present in the environment. The regulatory concentration limits for routine releases from existing nuclear facilities, which are derived from annual limits for intake (STA, 1988), can, however, be used to derive a common scale for all radionuclides of concern, allowing indirect comparisons to be made, as explained in detail in Takasu et al. (2000) and Miyahara et al. (2001).

4.3   Use of safety indicators for "What-if?" scenarios

There are a few "What-if" Scenarios in H12 in which radionuclides are not distributed within the surface environment in accordance with the assumptions of the Reference Biosphere. For these scenarios, indicators other than dose are the most appropriate.

Uplift and erosion

A "What-if" Scenario was considered in H12 in which the repository becomes exposed at the ground surface due to a prolonged period of uplift and erosion.

The safety indicator for this scenario was the rate at which radionuclides are released, expressed in terms of an equivalent release of ^{238}U, as the repository and its surrounding host rock are eroded at the surface. A yardstick is defined in terms of the rates at which radionuclides would be released by natural materials (granite or uranium ore) being eroded over the same area (the repository "footprint") at the same rate.

The equivalent release of ^{238}U from the repository in this scenario is comparable to the yardstick, showing that radiological perturbations are not insignificant in terms of natural phenomena, although the very long timescales required for exposure of the repository (at least some hundreds of thousands of years) must be born in mind when interpreting these findings.

Volcanic activity

Another "What-if?" Scenario was considered in which a major magmatic intrusion into the repository takes place one hundred thousand years after disposal, and all the radionuclides within the repository are released (siting criteria in Japan will ensure that the likelihood of such a scenario is extremely small).

The safety indicator for this scenario was the radionuclide release from the repository at the time of occurrence of this event, expressed in terms of an equivalent release of ^{238}U. The yardstick was the natural uranium contained in a Japanese volcano of average size (a volume of 4×10^{10} m^3, with an assumed uranium concentration of 1 ppm).

The equivalent release of ^{238}U from the repository in this extreme scenario is 5 orders of magnitude less than the yardstick. Although it must be remembered that the entire volume of a volcano is not discharged in a single eruption, it can nevertheless be inferred that the presence of the repository only slightly perturbs the radioactivity naturally released in an eruption.

4.4 Use of safety indicators to illustrate radionuclide containment

The long containment times of radionuclides within the multiple barriers provided by the disposal system means that most decay to insignificant levels before reaching the human environment. The decrease in the radiotoxicity of the contained radionuclides as a function of time was illustrated by means of a safety indicator termed the radiotoxicity index. The radiotoxicity index of a specific radionuclide within a specific system component has units of m^3 and is here defined as the inventory of that nuclide within the component (in Bq) divided by the maximum permissible activity concentration for that radionuclide outside a designated monitored area (in Bq m^{-3}) (STA, 1988). The "yardstick" was the radiotoxicity index of 5×10^{-5} km^3 granite (which corresponds roughly to the volume of rock above the area occupied per waste package) with 1 ppm natural uranium concentration.

Figure 4 shows the combined radiotoxicity index of all radionuclides in different system components as a function of time for the H12 Reference Case (one particular representation of the Base Scenario, defined in Section 5, below). The figure shows the substantial decay of radiotoxicity with time. The figure shows that radiotoxicity is contained predominantly within the engineered barriers of the repository at all calculated times. Only at times of around 10^6 to 10^7 years is a significant proportion of the total radiotoxicity present within the geosphere, by which time it has decayed by 5 orders of magnitude. The amount of radiotoxicity in the geosphere is always less than the natural radiotoxicity of 5×10^{-5} km^3 of granite. At no stage is a significant proportion of the radiotoxicity present in the biosphere.

Figure 4. The combined radiotoxicity index per waste package of all radionuclides in different system components as a function of time for the H12 Reference Case

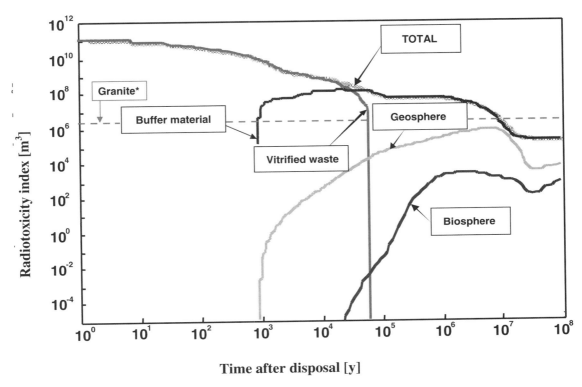

* 5×10^{-5} km^3 of granite with 1 ppm uranium concentration which corresponds roughly to the volume of rock above the occupied area per waste package.

5. Treatment of uncertainty in the evolution of the disposal system

The methodology for the treatment of uncertainty in the evolution of the disposal system, and, in particular, uncertainties in the occurrence, time of occurrence and degree of impact of features, events and processes (FEPs), has been developed further since H12. In the current methodology, these uncertainties are generally assessed by means of expert judgement. The detailed description of the methodology is beyond the scope of the present paper, but some key aspects are described below.

For each FEP, experts are first asked to judge *whether* a FEP should, or should not, be taken into account in assessing the radiological consequences of the system. The judgement is based on:

- whether or not the occurrence of the FEP at some time in the future is plausible – an expert might recommend the exclusion of a FEP on the grounds of implausibility;

- whether or not the FEP could have a significant impact on safety – an expert might recommend the exclusion of a FEP on the grounds of its irrelevance to safety; and

- whether relevant FEPs the occurrence of which is uncertain (though not implausible) are positive or detrimental in terms of their impact on safety – an expert might recommend exclusion of uncertain positive FEPs on the grounds of conservatism.

The views of the experts are then assessed to evaluate the degree to which there is consensus on inclusion or exclusion. For some FEPs there is likely to be unanimity as to whether or not they should be included. For others, there is likely to be a divergence of views. The categorisation of FEPs according to the degree of consensus on inclusion is illustrated by the y-axis of Figure 5.

Experts are also asked to judge *how* FEPs should be taken into account in assessing the radiological consequences of the system. The approach is to define different scenarios for the evolution of the system. There are three types of scenario: a single Base Scenario, a set of Perturbation Scenarios and a set of "What-if?" Scenarios. The judgement as to which scenario a FEP should be assigned to is based on:

- whether or not the FEP is part of the "design basis" of the system – FEPs that are part of the design basis are included in the Base Scenario; and

- the plausibility of the FEP – likely FEPs defining the expected evolution of the system are included in the Base Scenario, plausible (generally detrimental) FEPs that are nevertheless not considered likely are included in the Perturbation Scenarios, and speculative FEPs are included in "What-if?" Scenarios.

Again, the views of the experts are then assessed to evaluate the degree to which there is consensus on how FEPs should be included in scenarios. The colours in Figure 5 illustrate the categorisation of FEPs according to how they are included in scenarios.

Finally experts are asked to judge the approximate *time periods* over which included FEPs are likely to have an impact on the evolution of the system. Key periods are, for the EBS:

- the period up to about 10^3 years, corresponding to complete containment in the overpack;

- the period up to about 10^4 years, corresponding to slow release from the glass matrix; and

- the period between about 10^4 and 10^5 years, during which time there may be alteration of buffer and backfilling materials.

For the geological environment, key periods are:

- the period up to about 10^4 years, during which time significant climatic/sea level change is not expected; and

- the period between about 10^5 and 10^6 years, during which tectonics changes may be important.

These periods are illustrated by the x-axis of Figure 5.

The fate of radionuclides in the Base Scenario, Perturbation Scenarios and "What-if?" Scenarios are assessed using models and datasets, with conservative assumptions and parameter choices made where there is insufficient scientific understanding to support more realistic assumptions and choices.

6. Conclusions

Performance assessments seek to make arguments for safety in the presence of uncertainties that, in general, increase as a function of the timescale addressed by a safety assessment. This paper has discussed why that uncertainties in the evolution of the biosphere can be treated separately from uncertainties in the evolution of the EBS and the geosphere, i.e. its different role in the safety case and the high degree of decoupling of the evolution of the EBS and the geosphere from events and processes in the biosphere.

The paper has focused particularly on the role of safety indicators complementary to dose and risk. It has been shown that complementary safety indicators can be used:

- to give additional support to safety arguments based on dose calculations, by avoiding at least some of the assumptions inherent in these calculations,

- to provide safety arguments for a small number of "What-if" Scenarios for which the Reference Biosphere approach is inapplicable, and

- to illustrate containment, and the very high degree of radioactive decay that takes place within the repository and its immediate surroundings, instead of focussing solely on radionuclide releases.

Complementary safety indicators have been used in these ways in the H12 safety assessment to provide a range of arguments that together build a convincing safety case.

Based on the experience and lessons learnt form H12, a systematic methodology is being developed for deriving scenarios that takes into account how the importance of different FEPs varies over the course of time.

Figure 5. Examples of FEPs and their categorisation based on plausibility for inclusion into assessment scenarios

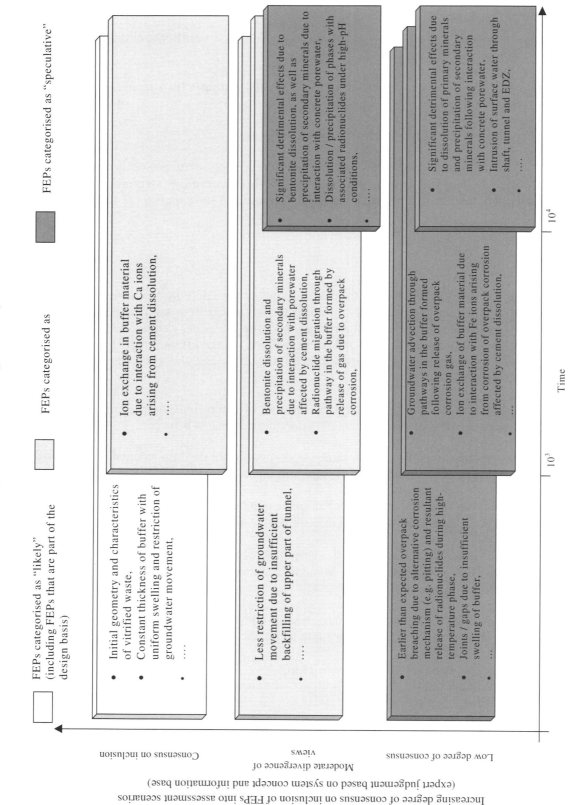

120

REFERENCES

[1]   AEC (1997): Atomic Energy Commission of Japan: Guidelines on Research and Development Relating to Geological Disposal of High-Level Radioactive Waste in Japan, Advisory Committee on Nuclear Fuel Cycle Backend Policy.

[2]   BIOMASS (1999): Long-Term Releases from Solid Waste Disposal Facilities: the Reference Biosphere Concept, BIOMASS Theme 1 Working Document: BIOMASS/T1/WD01, International Atomic Energy Agency, Vienna.

[3]   JNC (2000): H12: Project to Establish the Scientific and Technical Basis for HLW Disposal in Japan – Project Overview Report – JNC TN1410 2000-001.

[4]   Kametani, K. and Tomura, K. (1976): Concentrations of ^{226}Ra and ^{210}Pb in Tap Water, Well Water and Rainwater and Adsorption of ^{210}Pb on Soil, Radioisotopes, Vol.25, No.7, pp.38-40 (in Japanese).

[5]   Kametani, K., Matsumura, T. and Asada, M. (1991): An Analytical Method for Uranium and Investigation of ^{238}U and ^{234}U Concentration in River Waters, Radioisotopes, Vol.40, No.3, pp.26-29 (in Japanese).

[6]   Miyahara, K. Makino, H., Takasu, A. Naito, M., Umeki, H., Wakasugi, K. & Ishiguro, K. (2001): Application of Non-Dose/Risk Indicators for Confidence-Building in the H12 Safety Assessment, IAEA Specialist's Meeting to Resolve Issues Related to the Preparation of Safety Standards on the Geological Disposal of Radioactive Waste, June 16-18, 2001, Vienna, Austria.

[7]   Miyake, Y., Sugimura, Y. and Tsubota, H. (1964): Content of Uranium, Radium and Thorium in River Water in Japan, The Natural Radiation Environment, pp.219-225.

[8]   STA (1988): Science and Technology Agency: Notification No. 15 (in Japanese).

[9]   Takasu, A., Naito, M., Umeki, H. & Masuda, S. (2000): Application of Supplementary Safety Indicators for H12 Performance Assessment, MRS 2000, 24th International Symposium on the Scientific Basis for Nuclear Waste Management, August 27-31, 2000, Sydney, Australia.

[10] Tsumura, A. and Yamasaki, S. (1992): Direct Determination of Rare-earth Elements and Actinides in Fresh Water by Double-Focusing and High Resolution ICP-MS., Radioisotopes, 41, pp.185-192 (in Japanese).

[11] WHO (1998): Guidelines for Drinking-water Quality, World Health Organisation.

HANDLING OF TIMESCALES AND RELATED SAFETY INDICATORS

Lise Griffault and **Eric Fillion**
ANDRA, France

1.   Introduction

The fundamental objective of a nuclear waste repository is a long-term protection of man and environment. To conduct safety analyses over long period of times, up to the million years, is not a straightforward question. For example, the required demonstration of safety, with respect to the fundamental rules of safety (RFS III.2.f.), is out of the scope of any experiment. The safety analysis of a geological disposal system is usually addressed using qualitative and quantitative approaches. But in any case, the evaluation is made for specific time-span and those have to be carefully chosen and discussed in terms of uncertainties, especially for the long term.

At ANDRA, a tentative time fractionation has been conducted for geological disposal using two types of approaches:

- The first one is based on a "phenomenological" description of a series of situations dependent on the timescale. The tentative time fractionating of the repository has been proposed for the operational and post-closure phases using the thermal, hydraulic, mechanical and chemical (THMC) processes occurring within the system. The safety case can then be based on a series of simulations, where the objectives are to assess the level of safety that can be reached for various internal defects or external events (see paper from G. Ouzounian on Phenomenology Dependant Timescale, Part C).

- The second one, the purpose of this paper, proposes to handle timescale using supplementary "safety indicators". Quantitative "measure" of safety is generally obtained using "safety indicators" such as the conventional calculation of the "individual dose" (mSv/y) which means a calculation of a direct radiological impact on individuals. This individual dose evaluation is usually preferred to others because it is a direct safety indicator, widely used with a large international consensus and it can be compared with existing standards (ICRP). However, dose calculation is not entirely satisfying because it is burdened with uncertainties associated to the assumptions made as regards to critical group, human behaviour and climate change. As well, the probability of exposure is not taken into account in the calculation. The risk, number of health effects per year, bears more information than the dose. It also allows comparison with level of risk associated with other industrial activities or natural risks. But, the risk is a complex tool if it is not well explained and underlying problems can arise with the estimation of probabilities.

With an objective of confidence building, ANDRA is seeking the use of complementary safety indicators within the framework of safety assessment of a geological HLW repository. With regards to the definition of a safety indicator, which must provide a measure of safety, directly or indirectly, and can be compared to a judged acceptable value, supplementary safety indicators can be assessed with the objective of moving towards a qualitative approach. Beyond safety issues, safety indicators may also make a valuable contribution to communication, to decision making concerning the repository, to support technological choices relative to concept or to handle timescale.

Safety analyses commonly refers to time, such as transfer time through each barrier, time to reach the biosphere, decay time, time for alteration of waste packages or waste containers, time until the next glacial period, time of geosphere stability, etc. Still, a direct link between safety and time is difficult to assess. This paper attempts to put into perspective timescales and safety indicators. The possible use and interest of complementary indicators, such as activity, radiotoxicity, radionuclides fluxes and concentrations, in the context of the safety assessment of a geological HLW repository is presented and discussed with the objective to handle timescale.

A conventional definition of a safety indicator is that it must provides directly or indirectly a measure of safety [1]. The "safety" indicator is related to safety or capable of reflecting a level of safety. A required condition is that for each safety indicator must exist a criterion of comparison (or a reference value) to a limit judged acceptable.

The hierarchy of safety indicators presented by IAEA [1] was used as a guideline to review the use and interest of those potential indicators:

Table 1. **From the Hierarchy of safety indicator by IAEA [1]**

health effects ⇳ Risk ⇳ individual doses ⇳	Direct safety indicator can be compared to existing standard (and natural background)
Concentration in the environment ⇳	Comparison with natural radioactivity
flux entering biosphere ⇳ Flux through the different safety barriers ⇳	Comparison with flux of natural radionuclides
Waste	Indicator of radiotoxicity based on waste inventory

Review of potential safety indicator as a tool to handle timescale focused on dose estimation, activity, radiotoxicity, fluxes concentrations and time.

2.   Results

2.1.   *Dose estimation (mSv/y)*

As mentioned, conventional quantitative estimation is usually assessed using the individual dose standard which means a calculation of a direct radiological impact on individuals. For instance, for a normal situation, this dose has to be lower than 0.25 mSv/y according to the RFS.III.2.f. The advantage of such a quantitative dose estimation is its common usage and it benefits of a large international consensus. It is a direct safety indicator and there are existing standards (ICRP). But such quantitative estimation doesn't take into account the probability of exposure, it corresponds to a hypothetical situation, and it affects a hypothetical critical group, although its selection relies upon cautious and reasonable approach.

2.2.   *Activity (Bq)*

The evolution of the amount of radioactivity left in the disposal as a function of time is an easy way to show the radioactive decay occurring through the lifetime of the disposal. From the curve of total activity [Figure 1], it can be seen that a significant part of the radionuclides contained in the waste has decayed within the first 500 hundred years. Furthermore, from the same activity curve, discrete time intervals, usually 4, can be defined upon radioactive decay constant of radionuclides:

1.   From 0 to 1 000 y. : decay of the short lived radionuclides (^{137}Cs, ^{90}Sr).

2.   From 1 000 to 10 000 y. : decay of the middle lived radionuclides (^{241}Am, ^{238}Pu).

3.   From 10 000 to 100 000 y. : decay of the long lived radionuclides (^{14}C, ^{94}Nb).

4.   beyond 100 000 y. : decay of the very long lived radionuclides (^{129}I, ^{238}U). Activity   is not a direct safety indicator but it bears some informations, and it is quite easy to understand. Activity can easily be calculated as a function of time, see Figure 1. Activity also allows comparison with some natural systems, especially with the original uranium deposits used for the fuel. Comparisons have been made considering the time necessary for the waste to recover an activity equivalent to the one resulting from an uranium deposit. In that respect, activity has been and is still a interesting tool for communication to the public. It cannot be used for safety demonstration but it gives a rough idea of the decreasing potential danger of the stored radioactivity. One other advantage is that it relies upon minimal uncertainties, mostly related to waste inventory.

2.3.   *Radiotoxicity (Sv/g)*

As mentioned in the above chapter, activity is not a satisfactory indicator for safety evaluation of a deep disposal. Indicators of radiotoxicity have been defined using notion of dose factor (Sv/Bq) from the different radionuclides of the waste, in order to give a global toxicity per unit of volume of weight of material.

Figure 1. **Activity as a function of time, example of TSPA VA-US DoE, 1998 [2]**

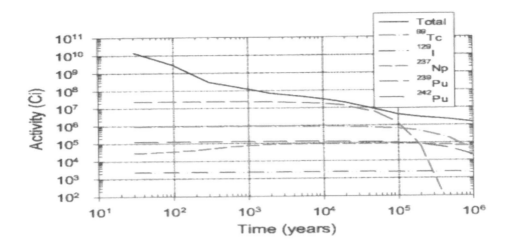

Radiotoxicity is often considered as an interesting tool for both internal and external use, such as communication to the public, but there is a still a limited usage of it even if it is quite easy to use and doesn't require complex software. A risk of confusion with dose exists if it is not properly explained. In that respect, quantitative results cannot (and should not) be compared to dose limits.

This indicator, when properly explained, is an interesting tool to illustrate the potential danger of the waste at any time during the life span of the repository. It can also be used in comparison with a natural material. Several examples of such an application exist as shown in Figures 3 and 4.

As a function of time, it allows to define discrete time intervals based on the radionuclides decay. Evolution of potential radiotoxicity by ingestion can be described for:

- Short period of time (<500y), corresponding to the decrease of short lived radionuclides.

- The period of about 5 000 y, corresponding to the decrease of middle lived radionuclides (^{241}Am).

- Period of about 10^5 y, corresponding to the decrease of long-lived radionuclides (^{239}Pu).

- Period over 10^7 y, corresponding to the decrease of very long-lived radionuclides.

Figure 2. **Example of the possible use of the radiotoxicity within the framework of the selection of concepts at ANDRA, France**

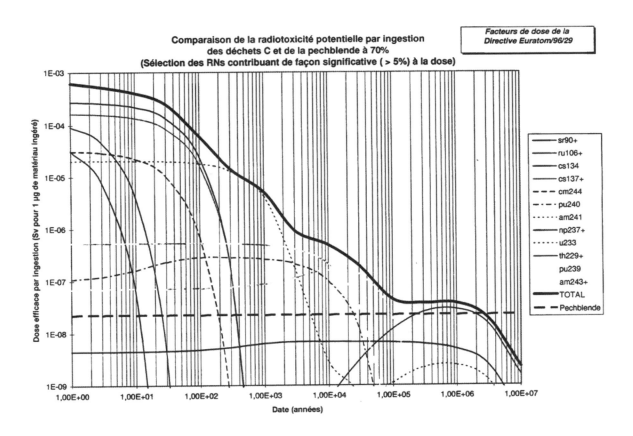

Figure 3. Example of the evolution of the radiotoxicity by ingestion – SKB (SR97), [3]

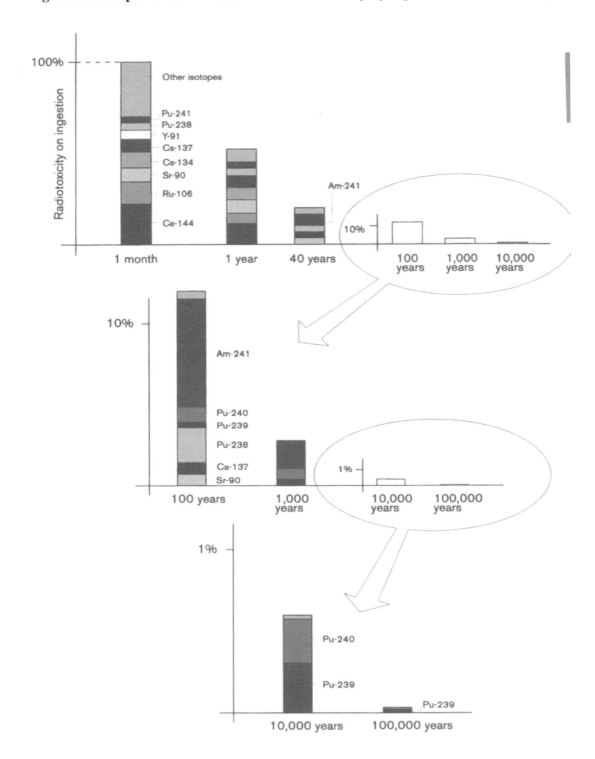

2.4.   *Activity fluxes (Bq/y, mol/y)*

Activity fluxes are not really safety indicators, rather they are performance indicators, especially when applied to barriers of deep disposal. They can provide a sound measure of barrier(s) efficiency by the evaluation the radionuclides fluxes attenuation (i.e. the ratio of input and output fluxes). Activity fluxes through engineered barrier or host rock illustrate, for example, the two major functions usually allocated to a deep disposal, i.e. "limitation" and "retardation".

Such an indicator has the advantage of being independent of the surface conditions as it relies only upon geosphere and its stability. The use of flux ratios also avoids referring to any natural analogue. Furthermore, it may help in guiding R&D on barriers, in comparing disposal concepts, or in comparing "confinement" properties.

Figure 4. **Example of the use of the activity fluxes in the DAIE Vienna, ANDRA 1996, [4]**

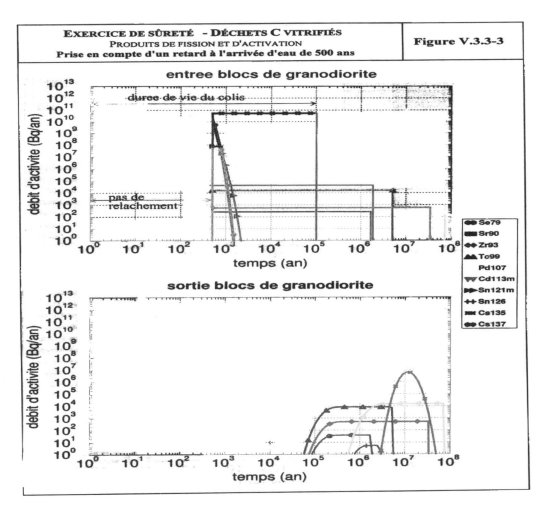

Activity fluxes can be defined for specific periods of time as they correspond to fluxes integration for a defined lapse of time. In that respect it can give a timescale, for instance 500 y for canister life, or 10^5 y for duration of the glass of type C waste. Example of applications are given by different organisations, one is given in figure 4.

2.5.   *Concentration in water (Bq/l)*

Concentration in water is the input data for dose calculation when transfer is in water (which is the main transfer of radioactivity the to surface in the case of deep disposal, with the subsequent consequence of exposition of the critical group by pumping the water).

Concentration, being widely used, is easy to understand for the public. Concentration represents a coherent approach with toxic substances. Indeed, toxic substances have "limits" given as a maximum allowed concentration in water. This standard usage of concentration makes it appropriate to consider concentration as a direct safety indicator in that case. Concentration also has the advantage of being experimentally measurable.

Concentration is also a good safety indicator in the sense of that it is independent of the surface conditions. Concentration is given in the aquifer, at the outlet, and it is, for instance, free of uncertainties linked to the definition of the critical group.

At the outlet, concentration can be a "measure" of the performance of the waste package. For instance, concentration values at the outlet can be compared to concentration of naturally occurring radionuclides. In such cases, the problem relies upon the reference values, and their local, regional, national, or international scale. Some national database exists through general survey but, it seems that an international consensus is not yet reached. Complementary information still needs to be collected.

The use of concentration is limited for artificial radionuclides because there is no "natural" reference value. It requires the development of methodologies to convert Bq/l to Sv/l, i.e. concentration of radiotoxicity. Several methodologies have been proposed for obtaining reference "acceptable" value. The first one described by Amiro and Zach at AECL, Canada [5, 6], focussed on the radiological impact, which must be lower than variations observed for natural background. It requires the collection of a set of data and statistical analysis in order to obtain the standard deviation. The acceptability is then defined by a radiological impact lower than the standard deviation. An example is given in figure 6. If the tritium concentration in the locality of the repository is 1Bq/l with a standard deviation of 0.3 Bq/l, then an increment of 0.1Bq/l (peak value) released from the disposal is acceptable.

Reference values for artificial radionuclides have also been proposed as a conversion of Bq/l to concentration of radiotoxicity based on the use of the dose factor by ingestion. Residual radiotoxicity for an artificial nuclides like ^{129}I is then compared to the residual radiotoxicity of a natural one like ^{238}U. The global reference level takes into account the sum of all radionuclides present in the natural system.

Even if concentration is restricted to transfer in water and is of limited use for artificial radionuclides, it remains a useful tool for engineers of repository conceptions.

Figure 5. **Adapted from the method developed at AECL (Canada), [5,6]**

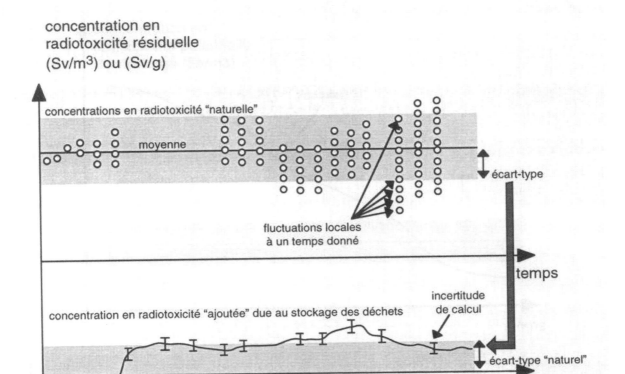

- Calculated amount of a released radionuclide is similar to the one naturally observed for the same element

- Calculated amount of a released radionuclides remains in the range of natural variations of values for the same element. The method was used by AECL (Canada). The calculated value is compared to an environmental increment, which is equal to the sigma value of the overall measured concentrations (Figure 6).

2.6.   Time (y)

Notion of timescales can be obtained through the following "specific times" even if they are not really safety indicators: 1) transfer time through each barrier, 2) time to reach the biosphere, 3) decay time, 4) radionuclides half-lives, 5) time for alteration of waste package or waste containers, 6) time until glacial period, 7) time of geosphere stability.

3.   Conclusions

This review emphasised the interest of complementary indicators. Dose and risk are quite complex quantities, which force researchers to seek form of simplicity more understandable to communicate to interested parties (regulator, local authorities, general public, the engineering community, etc.).

From this review, it can be shown that complementary safety indicators are of multipurpose use:

- They may be helpful in evaluating long term uncertainties. The safety assessment requires a set of hypotheses that are burdened with uncertainties such as surface conditions, human behaviour, climate change, radionuclides transport modelling, exposure scenario, etc.

- Complementary safety indicators can enhance confidence building, they can certainly be helpful in demonstrating our ability to evaluate safety in different manners (multiple lines of reasoning).

- Complementary safety indicators can be used as tools for the decision process.

- They are useful for comparison with natural analogues (concentrations), other industrial and social activities (risks).

- They also give some time perspective. From each of the indicators that was presented and discussed, it is possible to define some discrete time intervals:

  - *Activity (Bq):* definition of time intervals based on the radioactive decay of radionuclides (10^3, 10^4, 10^5, $>10^5$ years).

  - *Potential Radiotoxicity (Sv/g):* illustrates the potential danger at an instant during the life span of the repository.

  - *Fluxes (Bq/y, mol/y):* lifetime duration, time for alteration of waste, transfer time through each barrier, time to reach the biosphere.

  - *Concentrations (Bq/l, mg/l, Bq/kg):* useful tool to provide perspective with familiar quantities (natural background, potable water standards) it is valid for a relatively long period of time.

REFERENCES

[1] Safety Indicators in Different Time Frames for the Safety Assessment of Underground Radioactive Waste Repositories, IAEA TECDOC 767, October 1994.

[2] Total System Performance Assessment – Volume 3: Viability Assessment (TSPA-VA) – US Department of Energy – Office of Civilian Radioactive Waste Management – DOE/RW-0508, December 1998.

[3] Deep Repository for Spent Nuclear Fuel – SR97 – Post-closure safety – Technical report TR-99-06-SKB, November 1999.

[4] Demande d'Autorisation d'Installation et d'Exploitation – Laboratoire de recherches souterrain de la Vienne – Vol. 2 : Mémoire, Chapitre 5, Analyse de sûreté du stockage et objectifs d'études, ANDRA, 1996.

[5] Amiro B.D., Zach R., *A method to assess environmental acceptability of releases of radionuclides from nuclear facilities*, Environmental International, Vol.19, pp 341-358, Pergamon Press Ltd.

[6] Amiro B.D. *Protection of the environment from nuclear fuel waste radionuclides: a framework using environmental increments*, The Science of the Total Environment, Vol. 128 pp. 157-189, Elsevier Science Publishers, 1992.

TIMESCALES IN THE LONG-TERM SAFETY ASSESSMENT OF THE MORSLEBEN REPOSITORY, GERMANY

Matthias Ranft and Jürgen Wollrath
Federal Office for Radiation Protection (BfS), Germany

Abstract

The Morsleben repository for short-lived low- and intermediate-level radioactive waste is located in a former rock salt and potash mine in a salt structure of the Aller valley fault zone in NE Germany. Until the end of the operational phase a waste volume of about 37 000 m^3 with a total activity of approx. $4.5 \cdot 10^{14}$ Bq had been disposed of. The closure of the repository has still to be licensed. Therefore, a sealing concept has been developed which provides for an extensive backfill of the remaining underground openings, and new safety assessments have been performed.

The examination of the Morsleben site and its surroundings enables a comprehensive determination and assessment of the potential hazards compromising the natural barriers around the repository through geological processes and water inflow from surrounding geological units. Most important for the safety of the repository is the limitation of water inflow into the salt structure by the overlying tight caprock.

The paper introduces the geological situation at the repository site and the safety concept. As a result of the scenario analysis the derivation of four different timescales caused by the geological situation and resulting from the backfill concept is presented. One additional time frame is related to the possible hazard of the radioactive waste emplaced in the Morsleben repository.

1. Introduction

The Morsleben radioactive waste repository (Endlager für radioaktive Abfälle Morsleben, ERAM) is located in the Federal State of Saxony-Anhalt between the cities of Braunschweig and Magdeburg. It has been in operation since 1971 as a repository for short-lived low- and intermediate-level radioactive waste. Until the end of the operational phase in 1998 a waste volume of about 37 000 m^3 with a total activity of approx. $4.5 \cdot 10^{14}$ Bq had been disposed of.

The repository is located in a former rock salt and potash mine in the salt structure of the Aller valley fault zone which started salt production in 1900 and ended production in 1969. The mine with its widespread system of drifts, cavities, and blind shafts extends over a length of 5.5 km in NW-SE direction and 1.4 km at the maximum perpendicular. The 525 m deep Bartensleben shaft connects 4 floors in various levels between 386 m and 596 m. Due to rock salt and potash production, many cavities exist in this former mine with dimensions of up to 100 m in length, 30 m in width and in height, amounting to a volume of about $8 \cdot 10^6$ m^3 of underground openings (Figure 1). After backfilling during salt production an open volume of more than $5 \cdot 10^6$ m^3 remains. Radioactive waste emplacement took place on or below the fourth level mainly in the Southern Field, the Western Field and the Eastern Field and to a minor extent in the Northern Field and the central part of the Bartensleben part of the mine. The backfill and sealing concept which has been developed for the closure of the repository provides for an extensive backfill of all remaining underground openings in order to meet the safety goals. Nevertheless, an open space of some $1\cdot10^6\,\mathrm{m}^3$ will remain after backfilling.

Figure 1. **Sketch of the mine openings**

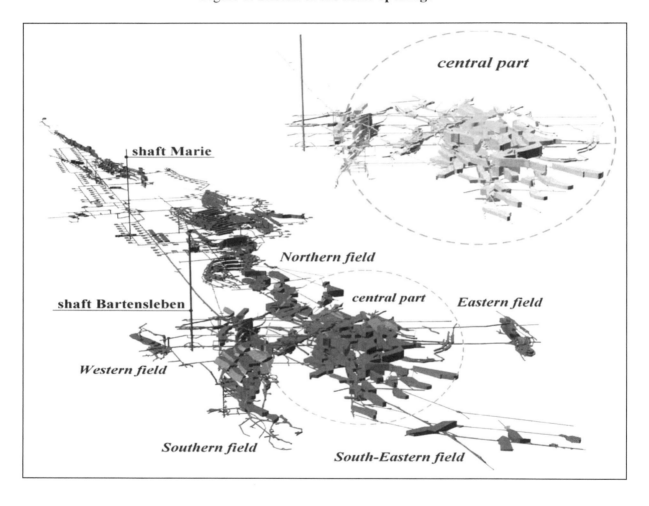

The license for operating the repository originates from the former German Democratic Republic and does not include the license for the closure of the repository. Therefore, according to the Atomic Energy Act (Atomgesetz, AtG, [2]) a license application for the closure of the repository has to be prepared by the Federal Office for Radiation Protection (Bundesamt für Strahlenschutz, BfS) who became the responsible operator of the repository after the re-unification of Germany in 1990. Based on the recently developed backfilling and sealing concept [4] new safety assessments have been performed aiming at the demonstration of the safe containment of radioactive waste.

As boundary conditions for the license application the German regulation neither stipulates that a subdivision of the safety assessment into different time frames is necessary nor provides a cut-off time beyond which no safety assessment has to be performed. The safety assessment has to fulfil the Safety Criteria for the Final Disposal of Radioactive Waste in a Mine [7] which introduce the

0.3 mSv/yr annual effective dose for an individual safety target. These Safety Criteria were set up in 1983 and are recently being revised to account for the international development since that time.

2. Geological situation

The approximately 50 km long Aller valley fault zone is a generally SE-NW striking major structural element of the NE German Subhercynian basin. It is between 1.5 km and 2 km wide and separates the Weferlingen Triassic block in the NE from the Lappwald block in the SW. The Aller valley salt structure is formed in the geological formation of Zechstein (Upper Permian) by intrusion into this fault zone (see Figure 2 and [1]). This intrusion was initiated during Buntsandstein times by migration of salt from the Weferlingen Triassic block and Lappwald block into the Aller valley fault zone. The salt formation is between 350 m and 550 m thick and consists of folded rock salt, potash salt and anhydrite stratifications. The potentially fractured anhydrite stratifications (Main Anhydrit) on the one hand stabilise the salt structure geomechanically and lead thus to a significantly low convergence of the underground openings in comparison to other salt structures and on the other hand represent a potential flow path for solution entering the salt structure.

Due to the rise of the salt structure during Late Cretaceous a thick caprock developed comprising Zechstein siliciclastic rocks, gypsum and anhydrite blocks which were formed by subrosion. This caprock is up to 220 m thick and forms an arched structure in the middle of the Aller valley fault zone. The caprock generally isolates the underlying salt structure from the aquifer system in the Cretaceous and Quaternary sediments. In the caprock the anhydrite blocks are associated with the siliciclastic and carbonatic layers of Deckanhydrit, Grauer Salzton, Leinekarbonat (DGL layer) which form a possibly more permeable feature within the tight caprock.

The central part of the caprock is overlain by unconsolidated rocks of cretaceous and quaternary age forming the main aquifer system in the Aller valley fault zone whereas the NE part is overlain by low permeability Middle Triassic mud stones and sand stones. The sequence of aquifers and aquitards in the Weferlingen Triassic block and in the Lappwald block drive the groundwater movement in the Aller valley fault zone which is directed mainly perpendicular to the valley and discharging into the Aller river [3].

3. Safety concept

As the release of radionuclides from the repository into the biosphere could not be excluded totally, it is limited according to German radiation protection regulations [5]. The safety objective is that the radiological impact on future generations should not exceed an annual effective dose of 0.3 mSv/yr per individual. In the German regulations the adherence to this safety objective is not limited to a certain time frame. Supplementary to these radiological safety objective conventional safety objectives like limitation of subsidence of the ground surface and protection of groundwater have to be taken into account.

Figure 2. **Geological cross section perpendicular to the Aller valley**

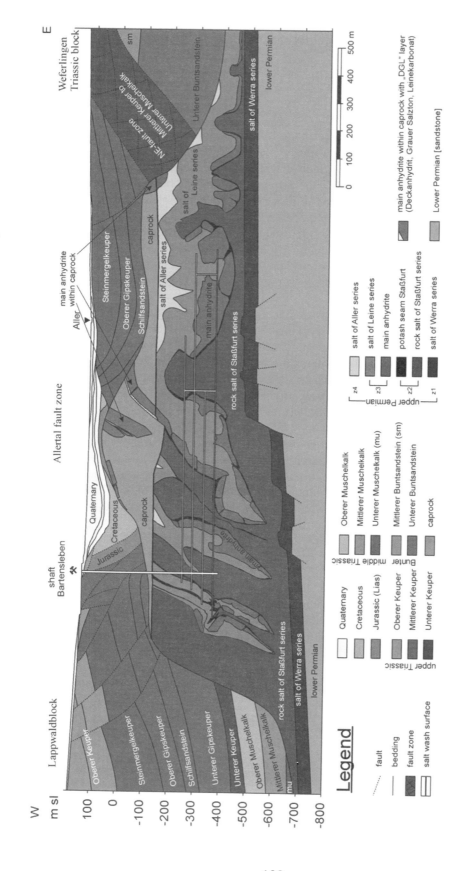

138

To achieve these safety objectives several closure concepts were considered. The closure concept which has finally evolved is based on a comprehensive backfilling of all mine openings and aims at the isolation and retention of the radionuclides in the disposal areas. The purpose of backfilling is:

- to preserve the integrity of the salt barrier by keeping the remaining openings mechanically stable and reducing the creep behaviour of the salt;

- to minimise the remaining open volume and to reduce the development of new cavities by solution of salt and potash in case of inflow of brine or water into the remaining openings; and

- to ensure a hydraulic resistance to mitigate or delay the flow of solution through possible pathways into the disposal areas and the flow of contaminated solution from the disposal areas to the biosphere.

The quality of the backfilling is divided into different categories according to its required function. The highest quality backfilling is required in the access drifts to the disposal areas where it should act as a hydraulic barrier to prevent solution inflow into the disposal areas. This engineered barrier is designed to have a permeability of 10^{-18} m^2 initially.

In case of solution inflow into the remaining mine openings a number of different processes firstly limit the amount of solution moving towards the disposal areas and finally limit and delay the transport of radionuclides from the disposal areas to the biosphere (see Figure 3, [6]).

Figure 3. **Important processes considered in the safety assessment**

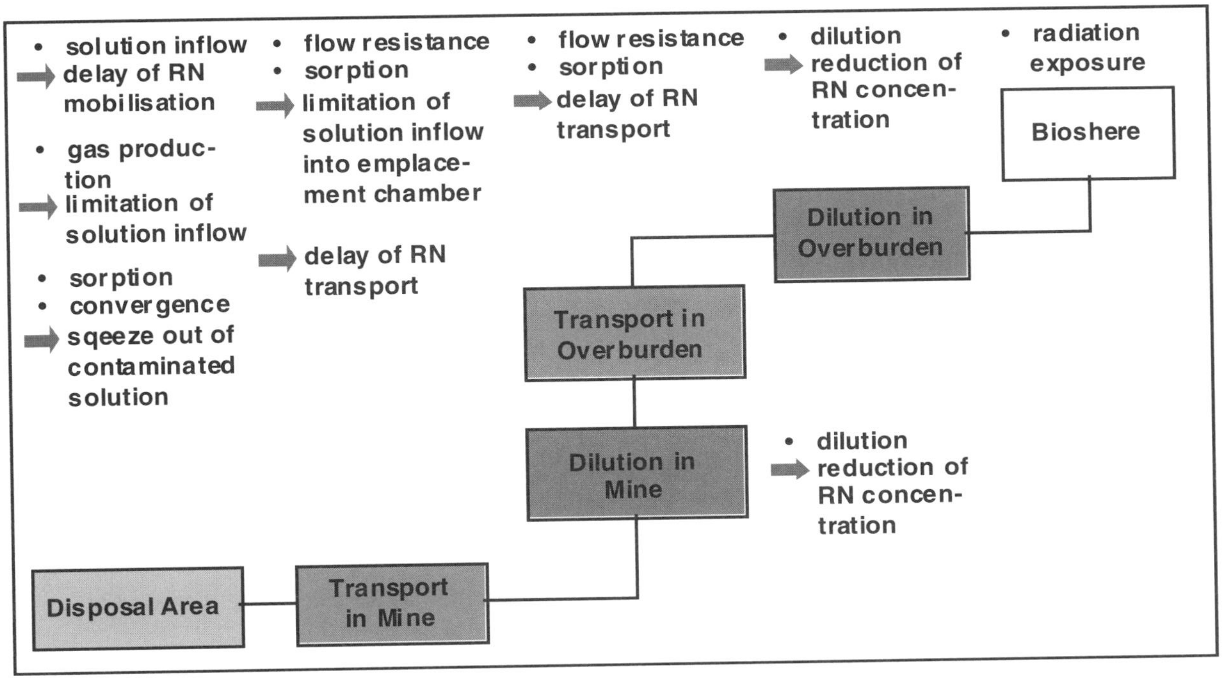

4.   Scenario analysis

The inflow of brine or water is the central scenario for the long-term safety assessment of the repository because it provides the transport medium for the radionuclides disposed of to get access to the biosphere. Currently five occurrences of brine inflow into mine openings are known, two of which are relevant due to possible connections to the surface rock. The main brine inflow is sited in the Marie part of the mine in a potash seam some 40 m below the top of the salt structure and was activated in 1907. With about 10 m^3/yr the volume of intruding brine has been constant over a long period of time. It is sealed from the main mine openings by a dam of masonry. The other occurrence is associated with a mine opening at the first level in the central part of the Bartensleben part of the mine. Above this mine opening anhydrite blocks in proximity to the DGL layer in the caprock are potentially conductive. The volume of brine intruding here amounts to about 1 m^3/yr.

Because of the importance of solution inflow into the disposal areas for repository safety, the scenario analysis aims at identifying the combination of potential locations from which brine or water originates with potential flow paths in the geosphere and in the salt structure. The main water reservoir is formed by the Cretaceous aquifers above the caprock in the Aller valley fault zone. Recently this reservoir is connected to the salt structure via the lowly permeable DGL layer in the practically impermeable caprock. At later times the hydraulic behaviour of the caprock might change due to re-activation of fractures caused by the change of tectonic stresses and the whole caprock might get more permeable. Flow paths in the salt structure are associated with the possibly fractured anhydrite blocks and with leaky rock salt mainly caused by the excavation of mine openings (EDZ). Having reached a mine opening at one of the upper levels the solution might migrate along the mine openings until it will reach the engineered barriers dividing the disposal areas from the remaining parts of the mine. To enter the disposal areas the solution must finally penetrate the engineered barriers. Due to gas production in the disposal areas and creep of rock salt the contaminated solution will then be squeezed out of the disposal areas and will migrate back to the biosphere taking similar flow paths.

At later times the climatic development leads to a cooling with the occurrence of permafrost which could modify the groundwater flow regime in the Aller valley fault zone. In the case of the development of an ice cover the present hydro-geological model looses its validity when the hydraulic conductivities of the aquifer system and the corresponding driving forces change substantially.

5.   Derivation of timescales

The scenario analysis identified the hydraulic behaviour of the caprock and the engineered barriers crucial for the safety of the Morsleben repository. Therefore, the derivation of time frames relevant for the safety assessment of the Morsleben repository has to account for this. Generally, relevant time frames could be subdivided into the three categories:

- Time frames resulting from site characteristics;

- Time frames associated to the behaviour of the engineered barriers; and

- Time frames representing the hazard potential of the radioactive waste disposed of.

The examination of the Morsleben site allows the subdivision of site characteristics into three conditions, representing different hydraulic situations. Each condition is associated with a certain time frame:

Condition A, representing the situation at the beginning of the post-closure phase: The Cretaceous aquifers above the caprock in the Aller valley fault zone are connected to the salt structure via the lowly permeable DGL layer in the practically impermeable caprock. Flow paths in the salt structure are associated with the possibly fractured anhydrite blocks and with leaky rock salt mainly caused by the excavation of mine openings (see Figure 4). Having reached a mine opening at one of the upper levels the solution might migrate along the mine openings or their EDZ until it will reach the engineered barriers dividing the disposal areas from the remaining parts of the mine. To enter the disposal areas the solution must finally penetrate the engineered barriers. Due to gas production in the disposal areas and creeping of salt the contaminated solution will then be squeezed out of the disposal areas and will migrate back to the biosphere taking similar flow paths.

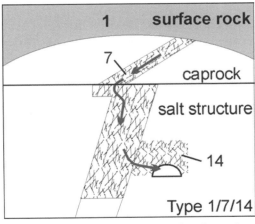

Figure 4. **Flow paths of solution into the salt structure at condition A**

The amount of solution entering the salt structure is limited by the hydraulic characteristics of the DGL layer. Based on the hydro-geological situation of the surface rock of the salt structure and its surroundings model calculations determined the maximum solution inflow into the salt structure to some 100 m^3/yr. Taking into account the initially open space in the salt structure of some $1 \cdot 10^6$ m^3, the fact that chemical reactions of the intruding solution with the different salt minerals might create new open space, and the creep behaviour of the salt which tends to reduce the open space the water inflow limited by the hydraulic behaviour of the DGL layer leads to a time frame of some 1 000 years. This time frame is needed to fill all mine openings before contaminated solution could be squeezed out of the salt structure as a result of the creep behaviour of salt and production of gas in the disposal areas.

Condition B, representing the situation after fracturing of the caprock: The Cretaceous aquifers above the caprock in the Aller valley fault zone are connected to the salt structure via fractured caprock. Flow paths in the salt structure are associated with the possibly fractured anhydrite blocks and with leaky rock salt connecting the mine openings to the salt leaching surface (see Figure 5). Having reached a mine opening at one of the upper levels the solution might migrate along the mine openings until it will reach the engineered barriers dividing the disposal areas from the remaining parts of the mine. To enter the disposal areas the solution must finally penetrate the engineered barriers. Due to gas production in the disposal areas and creep of salt the contaminated solution will then be squeezed out of the disposal areas and will migrate back to the biosphere taking similar flow paths.

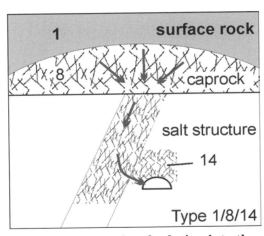

Figure 5. **Flow paths of solution into the salt structure at condition B**

For the creation of new fractures or the hydraulic re-activation of pre existing fractures in the caprock the stress field in the Aller valley fault zone must change substantially. Due to the past tectonic development of this region this could be excluded for the next few 10 000 years. After about 30 000 years, however, the stress field might have been changed so that the creation of fractures or the hydraulic re-activation of pre-existing fractures in the caprock has to be assumed. As a result the hydraulic conductivity of the caprock might increase and result in an immediate flooding of the remaining open space in the salt structure if there is any, and in a change of the retention behaviour of the caprock because flow paths through the clay-rich sorbing DGL layer are substituted by flow paths

through fractures without sorption capacity. This leads to a time frame of 30 000 years where the present hydro-geological model (condition A model) is valid and a time frame beyond 30 00 years where an alternative hydro-geological model (condition B model) has to be used in the safety assessment.

Condition C, representing the situation after development of an ice cover: At later times the climatic development might lead to a cooling with the occurrence of permafrost conditions with increased subrosion and finally the generation of an ice cover which could modify the groundwater flow regime in the Aller valley fault zone because of a substantial change of the hydraulic conductivities of the aquifer system and the cor-responding driving forces (see Figure 6).

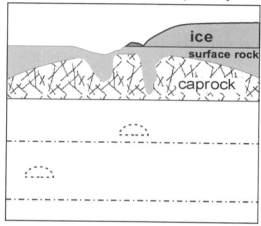

Figure 6. **Sketch of the hydro-geological situation at condition C**

The forecast of the climatic development leads to cooling phases after some 60 000 years and some 100 000 years. The hydro-geological model based on the present geometry looses its validity after approximately 150 000 years when an ice cover might have generated in the region of the Morsleben site. This leads to a time frame of between 30 000 years and 150 000 years, and the condition B model has to be applied in the safety assessment. For the time frame starting at 150 000 years no serious assumptions about the hydraulic behaviour of the surface rock of the salt structure could be made.

Time frame associated with the behaviour of the engineered barriers: Apart from the time frames resulting from the site characteristics an other time frame is associated with the characteristics of the engineered barriers which isolate the disposal areas from the remaining parts of the mine. These engineered barriers will be constructed in the access drifts to the disposal areas using a mixture of crushed salt, cement, filter ashes from conventional power plants, and water known as salt concrete. Laboratory experiments indicate that it will attain an initial hydraulic conductivity of at least 10^{-18} m^2. Brine migration through the barriers might degrade their hydraulic behaviour due to chemical reactions, however. This could not be avoided because no material has been known that is chemically stable to all possible brines getting in contact with the barriers. The chemical constitution of the brine depends on its flow path and the type and amount of salt minerals (e.g. rock salt, potash) associated with that flow path and could not be pre-determined due to the geometrical complexity of the mine and the salt structure.

Laboratory experiments and model calculations indicate that the hydraulic conductivity of the engineered barriers will increase depending on the amount and chemical constitution of the brine migrating through it. Penetration of the engineered barriers starts when a significant hydraulic gradient has established between the disposal areas and the remaining parts of the mine. This is connected to the site characteristics (condition A above) and requires that the mine openings are flooded. Due to the initially low permeability of the engineered barriers model calculations based on laboratory experiments show that in the worst case constitution of the brine they will keep their function for at least some 5 000 years after having got in contact with that brine. Afterwards they will no longer act as hydraulic barriers between the disposal areas and the remaining parts of the mine. This evolution of the barrier behaviour of the engineered barriers leads to a time frame of 5 000 years, starting when all mine openings are flooded.

Time frame representing the potential hazard of the radioactive waste disposed of: In addition to the comprehensive model calculations performed during the safety assessment for the Morsleben repository the potential hazard of the radioactive waste disposed of is evaluated by simple dilution considerations. It is assumed that all radionuclides are dissolved in a brine volume of 250 000 m^3 representing the volume of the Western Field, the Southern Field, the Eastern Field, and a small part of the remaining parts of the mine representing the flow path from the disposal areas to the top of the salt structure. This contaminated brine is squeezed out of the salt structure by convergence of the salt with a rate of 10 m^3/yr and directly diluted with 25 000 m^3/yr representing the flow rate in the Cretaceous aquifer. Subsequently, the radiation exposure of a person using this contaminated groundwater is calculated.

After 10 000 years the calculated dose rate for that person amounts to less than 1 mSv/yr which compares to the dose rate of 2.4 mSv/yr resulting from natural radiation in Germany. Therefore, this simple dilution considerations demonstrate that the potential hazard of the Morsleben repository could be characterised by a time frame of 10 000 years.

6. Summary and final remarks

The examination of the Morsleben site and its surroundings enables a comprehensive determination and assessment of the potential hazards compromising the natural barriers of the repository through geological processes and water inflow from surrounding geological units. Most important for the safety of the repository is the limitation of water inflow into the salt structure by the tight caprock.

Despite the fact that no subdivisions into different time frames are provided for in the German regulations, in the scenario analyses and the safety assessment, a number of time frames were identified which influence the further behaviour of the repository system:

- As long as the caprock keeps its present hydraulic condition the amount of brine or water inflow is limited to some 100 m^3/yr at maximum if an inflow occurs at early times. In conjunction with the amount of open space remaining in the mine and the creep behaviour of the salt which tends to reduce these openings the limited water inflow leads to a time frame of some 1 000 years which is needed to fill the mine openings before contaminated solution could be squeezed out of the salt structure.

- Due to tectonic reasons the hydraulic behaviour of the caprock can change after some 30 000 years by the creation of additional conductivity. Then, the caprock no longer limits the rate of water inflow into the openings present in the salt structure at that time. This results in a nearly instantaneous flooding of the remaining openings if there are any and an immediate begin of solution being squeezed out of the salt structure. Additionally, the radionuclide retention behaviour of the caprock is changed.

- The climatic development leads after some 60 000 years and some 100 000 years to a cooling with the occurrence of permafrost which could modify the groundwater flow regime in the Aller valley fault zone. The present hydro-geological model looses its validity after approximately 150 000 years when the hydraulic conductivities of the aquifer system and the corresponding driving forces may change substantially.

- Another important time frame is related to the development of the hydraulic conductivity of the engineered barriers which divide the disposal areas from the remaining parts of the mine. Due to the fact that the material of the engineered barriers could be effected by the brine established in the mine due to possible contact with different salt minerals the initial hydraulic conductivity might be increased depending on the amount of brine migrating through it. Even in the worst-case constitution of the brine they will keep their function for at least some 5 000 years after having got in contact with that brine.

- A time frame connected to the potential hazard of the radioactive waste disposed of is evaluated by simple dilution considerations. After 10 000 years the calculated dose rate for a person using the contaminated groundwater amounts to less than 1 mSv/yr which compares to the dose rate of 2.4 mSv/yr resulting from natural radiation in Germany. This 10 000-year time frame demonstrates that longer time frames and at least the time frame related to the climatic development at the Morsleben site are not significant for the safety of the Morsleben repository.

REFERENCES

[1] Albrecht, H.; Balzer, D.; Käbel, H.; Langkutsch, U.; Lotsch, D.; Putscher, S.; Ziermann, H.; Geological Setting of the Morsleben Radioactive Waste Repository, with Emphasis on Hydrogeological Modelling. In: Slate, S.; Feizollahi, F.; Creer, J.(Eds.): Proc. ASME 5th Int. Conf. on Radioactive Waste Management and Environmental Remediation (ICEM '95), Vol. 1, Berlin, 3-8 September 1995. New York: The American Society of Mechanical Engineers, 1995, pp. 1289-1293.

[2] Atomic Energy Act (Gesetz über die friedliche Verwendung der Kernenergie und den Schutz gegen ihre Gefahren – Atomgesetz, AtG). BGBl. part I, 2000, p. 1960.

[3] Ehrminger, B.; Genter, M.; Klemenz, W.; Wollrath, J.: Calibration of the 3D-Groundwater Model of the Post-Permian Sedimentary Cover at the Morsleben Radioactive Waste Repository Site. In: Stauffer, F.; Kinzelbach, W.; Kovar, K.; Hoehn, E.: Calibration and Reliability in Groundwater Modelling – Coping with Uncertainty. Proc. ModelCARE 99 Conf., Zurich, 20-23 September 1999. IAHS Publication No. 265. Wallingford: IAHS Press, 2000, pp. 177-184.

[4] Köster, R.; Maiwald-Rietmann, H.-U.; Laske, D.: Development of Backfill Materials in an Underground Salt Repository. Transactions 6th Int. Workshop on Design and Construction of Final Repositories "Backfilling in Radioactive Waste Disposal", Brussels, 11-13 March 2002, ONDRAF/NIRAS, Brussels, 2002.

[5] Radiation Protection Ordinance (Verordnung über den Schutz vor Schäden durch ionisierende Strahlen – Strahlenschutzverodnung, StrlSchV). BGBl. part I, 2001, p. 1714.

[6] Resele, G.; Oswald, S.; Jaquet, O.; Wollrath, J.: Morsleben Nuclear Waste Repository – Probabilistic Safety Assessment for the Concept of Extensive Backfill. In: A. Roth (Ed.): Proc. DisTec 2000 – Disposal Technologies and Concepts, Int. Conf. on Radioactive Waste Disposal, Berlin, 4-6 September 2000. Hamburg: Kontec, 2000, pp. 573-578.

[7] Safety Criteria for the Final Disposal of Radioactive Waste in a Mine (Sicherheitskriterien für die Endlagerung radioaktiver Abfälle in einem Bergwerk). GMBl. 1983, p. 220.

LONG TIMESCALES, LOW RISKS: RATIONAL CONTAINMENT OBJECTIVES THAT ACCOUNT FOR ETHICS, RESOURCES, FEASIBILITY AND PUBLIC EXPECTATIONS – SOME THOUGHTS TO PROVOKE DISCUSSION

Neil A. Chapman
Nagra, Switzerland

Abstract

This paper discusses a range of technical and non-technical factors related to long timescales for deep geological repositories. It is intended to provoke discussion between implementers, regulators and the public about realistic containment and protection objectives for long-lived wastes such as spent fuel and HLW. The ethical and practical aspects of providing protection are discussed, along with society's perceptions of hazard, protection and time. A proposal is made for a series of time-graded containment levels that reflect objectively achievable and ethically reasonable protection for future generations.

1.   Introduction

The future timescales that we find ourselves dealing with in assessing and presenting the safety of deep geological repositories are unprecedented in both engineering and decision-making experience. The combined community of physicists, nuclear engineers, earth scientists and radiological protection experts that has advocated geological disposal appears to have conspired, inadvertently, to convert what was originally a clear and logical concept for safely isolating wastes into something that critics would assert requires a degree of foresight not found in any other human endeavour: foresight that is inevitably shrouded with uncertainties, making it vulnerable to simple challenges. Implementers aim to guarantee doses that are always a small fraction of a variable natural background: hypothetical doses to hypothetical people, living ten or twenty times further into the future than the Ancient Empire of bronze age Egypt existed in the past.

Each discipline working in the field of geological disposal has added its own burden of independently sound, but only loosely compatible logic to the debate: half-lives are very long and must determine containment times; geological and environmental change is complex and must be accounted for fully when evaluating safety; we must protect anyone at any time in the future as if they were exposed to present-day nuclear industry practices. The result is that the nuclear establishment (both implementers and regulators) in some countries is being asked to make overly specific commitments about the distant future that it clearly cannot guarantee, and in so doing is expending considerable resources in order to avert risks, well beyond any normal human time horizon, that are arguably of no consequence. In all this, we also seem to have lost perspective in terms of how we deal with the risks associated with other natural and anthropogenic environmental hazards.

How should we respond to this situation? This paper looks at the ethical position of what we should hope to achieve in the near and far future when managing long-lived wastes and proposes a pragmatic restatement of the rationale for geological disposal, based on realistic containment and protection goals for different periods in the future.

To set the stage, it is useful to think about what motivates containment objectives in the context of the level of hazard posed by wastes at different times. Clearly, the greatest hazard with most wastes destined for geological disposal is associated with the early time period of a few hundred years. Here, when there is the potential for major impacts if people were to be exposed to the wastes, the design process aims to provide a containment system that gives maximum protection, and the need for compliance with radiological protection regulations should be at its most stringent. After this high hazard period, containment requirements need not be any more demanding and, in fact, become less rigorous with increasing time. At some point in time, containment can be argued to become unnecessary (and, at some later time, unachievable). When does the importance of the containment system decline to the former point?

To provide one possible anchor point in time, Figure 1 shows the progressive reduction in toxicity of spent fuel compared to that of an equivalent amount of natural uranium ore used to manufacture the fuel. The key message, whose significance will be discussed in the final section of this paper, is that there is a "back-to-nature crossover" point when the hazard of a spent fuel repository becomes similar to that of natural systems, albeit not necessarily systems found in similar environments or regions of the planned repositories. Depending on the assumptions used, this cross-over occurs after between one to a few hundred thousand years for both spent fuel and HLW. As will be discussed, it could provide a valid benchmark in time for the provision of protection: one beyond which both our ambitions and what society might expect of us change to a different gear.

Figure 1. **Relative radioactivity of typical spent fuel (a Swedish BWR fuel) as a function of time after discharge from the reactor, showing the early, dominant contribution of the fission and activation products. The sharp plunge in fission product activity between 100 and 1 000 years is largely a result of the decay of ^{90}Sr and ^{137}Cs, both with half-lives of about 30 years. After a few hundred years, the actinide elements become dominant. After a few hundred thousand years, the total activity of the fuel is similar to that of the uranium ore from which the fuel was produced (redrawn, after Hedin, 1997)**

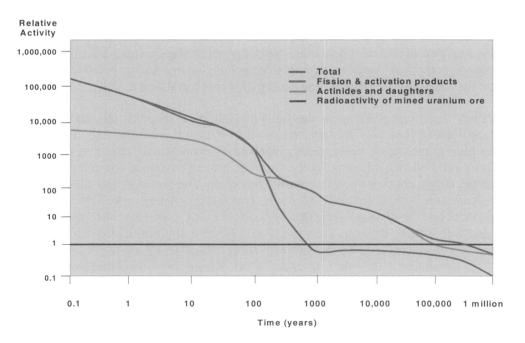

2.   Ethics and resources

The key ethical positions on long-term safety are, to some extent, contradictory. Primarily, it is generally stated that we plan to provide the same measure of safety for all future generations. This is sometimes called the "trustee" principle, whereby we must consider the interests of future generations, and is also linked to the "sustainability" principle, whereby we should endeavour not to pass on undue burdens to such generations. Second, those who are concerned about our knowledge on the safety of disposal exhort us to do nothing that could cause irreversible harm unless there is some compelling need. This is called the "precautionary" principle. Third, we should use our resources to provide for the needs of immediate generations, putting priority on near-term hazards rather than on hypothetical long-term hazards: the "chain of obligation" principle (NAPA, 1997).

There are obvious questions raised by these principles:

1. do the trustee and sustainability principles imply providing exactly the same style of radiological safety for all future times, or would broader standards related more to common natural exposures than to those applied to current nuclear industry practices be more appropriate: is it actually possible to state with confidence that we <u>can</u> provide the same style of safety in the far future?

2. at what level of predicted "harm" do we activate the precautionary principle: many activities of people cause a degree of irreversible harm, but below some level, that harm may be locally or regionally insignificant and, in the longest time frames, natural dispersal processes ensure that no harm that could conceivably arise from radioactive waste disposal is irreversible;

3. if we had serious limitations on resources, or even if we were just concerned about sensible use of available resources, could we ethically opt to protect current generations and forego unduly onerous provisions to protect future generations, provided we believed that any likely harm was small, uncertain to occur, and likely only to affect a few hypothetical individuals in the distant future?

The second question concerns the much-interpreted precautionary principle, whose current treatment highlights a key aspect of environmental ethics, which is that any decisions on a proposed activity are inevitably political, even though they may be conditioned by scientific studies of risk. The European Commission (EC, 2000) has provided advice on application of the precautionary principle that emphasises this and which notes, among many other recommendations, that "the appropriate response in a given situation is thus the result of a political decision, a function of the risk level that is "acceptable" to the society on which the risk is imposed". In the absence of knowledge about what would be acceptable to future generations, the easiest option is to assume that it would be the same as we "accept" today: unfortunately, society "accepts" different levels of risk for different activities.

An examination of social perspectives reinforces another key point that has been brought forcibly home to implementers in particular – we are not solving the waste problem in isolation, as "guardians" who know best. We are acting on behalf of society, so we also have to ask how society would like to weight ethical principles. Does society care about or is it indifferent to future generations? How would society like to spend its resources? Unfortunately, society has no common ethical or value-based view, and is seldom asked specific questions about how to divide and allocate a pool of resources so as to solve multiple "future" problems or to improve future conditions. Consequently, some people assume that there are sufficient resources available to protect all people completely, for all time. The scientific approach to this issue is to introduce optimisation of protection, expending resources only where risk can be reduced by reasonable expenditure of resources, and then only when some obvious benefit accrues from reducing risks. If risks are already insignificant, then it is not worth spending any resources to reduce them further.

Although risk tolerability and ALARP are neat concepts for allocating resources, they are used erratically or not at all by political decision makers and society. "Risk informed decision making" for technical matters is still largely an academic concept that falls down as soon as a decision-maker is faced with forcibly expressed, or just perceived potential public concerns. This has led to inappropriate measures being taken in cases where the objective risks are estimated to be trivial.

Trade-offs between the present benefits of alternative use of resources and the reduction of hypothetical far-future harm are never considered by either society or its political representatives: if they were, we would logically be spending no money at all on P&T research for radioactive waste management and a vast amount more on regional health immunisation programmes or natural hazard mitigation for communities in ecologically, climatically or tectonically sensitive areas. Making information more readily interpretable in terms of hypothetical *benefits* (of alternative uses of resources) as a function of time, rather than hypothetical *detriments* might be a way forward.

3.   Real public concerns about the future

Most people are seriously concerned about the safety of future generations no further than their grandchildren: less than a 100 year time frame. A recent study of public opinion in Japan, UK and Switzerland (Duncan, 2001) showed that 75-80% of people who were questioned thought only this far forward when considering the future welfare of themselves and their family, and more than 90% only looked as far as 500 years into the future. The latter time horizon was also cited by more than 90% of people when considering a wider social perspective: the future welfare of their township. 80-90% of Swiss and UK respondents asked about their timescale of concern for the global environment stopped at 1 000 years.

Given the short forward timescales with which people are concerned it is not surprising to hear that real worries are about whether a repository next door will irradiate people each time they drive past it; will it leak and poison their water; will it contaminate their community's land? This suggests that we should apply much effort to reassuring people about what we (quite perversely, in the public view) would call "short-term" safety. The most recent Eurobarometer survey of attitudes to radioactive waste in fifteen EU countries (INRA, 2002) showed that, overall, about 58% of those who had a view on living near a deep repository were most concerned with transport of waste to the site, leaks during operations or reduced property values. The remaining 42% were concerned about environmental impacts over the next hundreds or thousands of years (which the poll, interestingly, in the light of the comment above, did call "long-term" effects).

A need to focus on the immediate is reinforced when we consider our generally poor experience of prediction over years or decades. Experience of scientific prediction is that, within a person's lifetime, things generally work out differently to how we were told they would. People are sceptical about claims to predict the integrity of passive, man-made, engineered systems for more than a few decades into the future, unless constant maintenance is assured. Going back to the study by Duncan (2001), it was found that the majority of those questioned believed that waste could only be contained in a repository for 100 years. This suggests a rethink to the conventional way that we present safety assessments. Basing core arguments on timescales associated with natural isolation mechanisms may be much more digestible. It is interesting to note that the same survey found that only 2% in the UK and 14% in Switzerland believed that the waste could be contained in the rock "forever".

Being more concerned about the immediate future is not the same as stipulating a "cut-off" beyond which no evaluation of safety or performance is presented. Some regulations do not require compliance with "early period" dose/risk targets beyond 10 000 or 100 000 years. If these targets could be exceeded under reasonably likely conditions beyond this period, the cut-off approach is clearly non-sustainable. What this does imply is some form of tacit (but generally unstated) discounting of future harm – an issue over which views differ markedly and which is returned to in the final discussion.

4.   The perception of time

As there is often some form of temporal symmetry in an individual's recognition of past and future timescales (Duncan, 2001), it is useful to introduce this section on perception of time by looking at the balance between time projections of typical safety cases and what has happened over commensurate periods in the past. Four examples are given:

1.   Some disposal programmes countenance a measure of control over a repository site for around 300 years, possibly even leaving a repository open to allow for ease of retrievability. This period is presumably to be managed by national institutions. 300 years ago, about one half of today's European nations did not exist.

2.   The whole of recorded human history happened in the last 5 000 years: about the time some concepts expect their waste containers to last.

3.   Modern human beings are believed to have appeared in Africa perhaps 150 000 years ago: about the time it takes for spent fuel to reach the "cross-over" to radioactivity and toxicity levels similar to the original uranium ore.

4.   Modern human beings did not reach Europe until 40 000 years ago: in some deep clay formations, it takes water this long to move one metre.

Figure 2 illustrates some aspects of the human and natural environment during the last 150 000 years: a relevant period because it can be related to the natural cross-over time shown in Figure 1. The symmetric period in the past that matches the times with which people are concerned about the future is so short that it would not be visible on the scale shown for Figure 2.

These types of comparison of past and future times give mixed messages: different people will feel reassured ("who really cares about times beyond comprehension") or worried ("how can we say anything at all about the future"), emphasising that, although people are clearly more worried about the near future, perceptions of long timescales vary considerably.

5.   Long times and small risks: huge and tiny numbers

For a properly sited repository, relying on long-term stability of a hydrogeologically non-dynamic, isolated system as the main pillar of safety, we are dealing with minute, probably unidentifiable and unattributable potential impacts, unless someone digs the waste up again. The health impacts we calculate for the far future are stochastic, addressable only by statistical methods. There are widespread misconceptions about the meaning of the risks that are calculated. The numbers themselves are surrounded by a growing dispute over whether such low incremental doses have health significance that should be of any concern to society. Trying to make a convincing case to the public by matching huge numbers (hundreds of thousands of years) with tiny numbers (microsieverts) on the same diagram can present a real burden. Something conceptually simpler and more robust is required.

6. What protection should we aim to afford in the distant future?

Geological disposal is a conscious attempt to concentrate, contain and isolate hazardous material in a location that is well out of harm's way. It can be done, with reasonable expenditure of resources, in such a way as to have zero effect on the biosphere for many thousands of years. Achieving zero impact, even for a few hundred years, is already a major advance on any other management option and on what society achieves for any other persistent hazardous materials. It also seems to match society's expectation of who they want to protect and for how long. We should make much more of this exceptional ability to solve an environmental problem

But we can do better than that. A properly implemented solution at a good site can contain long-lived wastes such as spent fuel and HLW until they have decayed to levels of hazard commensurate with natural uranium ore deposits: a few hundred thousand years. It is completely unavoidable that "concentrate and contain", which provides such exceptional protection for much longer periods than society has any real interest, leaves the inevitable by-product of something akin to an ore body. In the tightest, highest containment rock formation, the longest-lived natural series radionuclides will remain exactly where they were placed, for geologically long periods of time. It could be argued that a rich uranium ore deposit was not originally present in this or that municipality, state, county, canton or country, but we are trying to achieve global solutions: by the time that the "ore deposit" can have any kind of impact in the environment, perhaps a million years in the future, such parochial considerations seem irrelevant.

Beyond this natural "cross-over" time there is a strong case, based on the parallel with nature, on society's real expectations and on sensible use of resources, for saying that we have done enough. There is no logical or ethical reason for trying to provide more protection than the population already has from Earth's natural radiation environment, in which it lives and evolves. It is a scientifically tenuous position to argue that additional protection (e.g. down to a few microsieverts of exposure) can be provided so far into the future and unreasonable to expect it, or to regulate for it.

What would this approach mean in terms of the protection provided by a deep repository? A system of simple "time-graded containment objectives" could be envisaged for the designers and siters of deep repositories, which would provide the following broad levels of protection:

Level 1. Zero impact: total containment of all activity in the repository for the period that is of concern to society: about 500 years.

Level 2. For the next one (or a few) hundred thousand years, any releases through natural mechanisms to give rise to doses that are below the range of natural background radiation.

Level 3. After this time, the hazard being equivalent to natural radiation hazards, there is no further containment objective: doses may be envisaged in the range of those from natural background radiation.

Figure 2. Aspects of the human and surface environment during the last 150 000 years

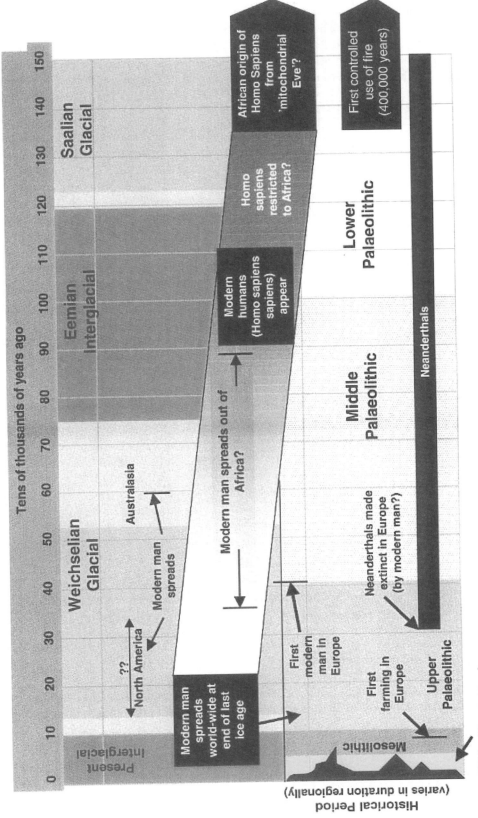

In more specific terms, Level 2 would be the period to which the spirit of current radiological protection principles could be applied. It is a period over which reasonable estimates of future system behaviour can be made. The performance measure appropriate to this period, and to the approach advocated here, would be to have reasonable expectation that any impacts (assuming the same biosphere as today) are less than about 10% of the world-wide variation in normal background radiation (excluding the highly variable radon contribution): a figure of around 0.3 mSv/a is appropriate. This would correspond to what ICRP, in its basis for discussion on new recommendations, is currently calling Band 3: Low Concern.

In the period beyond one (or a few) hundred thousand years (Level 3), it must be recognised and accepted that the potential exists for uranium ore deposits, or spent fuel or HLW repositories, to give rise locally to doses that are higher than the global average for natural radiation (~2.5 mSv/a), particularly if they are eventually eroded in the near-surface environment. However, as people exist today in many locations where doses are tens, even up to a hundred times (e.g. Ghiassi-nejad, et al, 2002), higher than the average, a repository is not providing, globally, a novel source of exposure.[1] Nevertheless, it might be expected that the eventual redistribution of residual radioactivity in the environment by erosion and other natural processes should be indistinguishable from regional variations in concentrations of natural terrestrial radioactivity in near-surface rocks, soils and waters: with "regional" taken in the broad sense of, for example, Europe or North America.

ICRP 81 (ICRP, 1998) notes that judgement is required in optimising protection, and says that the Commission's view is that, provided that the appropriate constraint for natural release processes is met, that reasonable measures have been taken to reduce the probability of intrusion and that sound engineering, technical and managerial principles have been followed, then radiological protection requirements can be considered satisfied. For the Level 2 period, the objective of the approach outlined here is the same, as is the natural release performance measure. The suggestion for Level 3 interprets ICRP 81 views on the progressively decreasing relevance of numerical performance measures with time much more broadly, but, it is believed pragmatically.

The intention of the approach suggested here is to make the very best of an excellent concept (geological disposal), to accept that with this excellence come unavoidable ultra-long-term implications, to do the very best that we can in engineering the solution and to provide optimum protection where society wants it most, and where we have the most responsibility. In doing this we are fulfilling our ethical responsibilities in the most practical way and going far beyond provisions for the future made in any other field of human endeavour.

REFERENCES

[1]   Duncan, I. J., 2001. Radioactive Waste: Risk, Reward, Space and Time Dynamics. Unpublished D.Phil thesis, University of Oxford, 401 pps.

[2]   EC, 2000. Communication from the Commission on the Precautionary Principle. Commission of the European Communities, Paper COM(2000)1, Brussels, 29 pps.

[3]   Ghiassi-nejad, M., Mortazavi, S. M. J., Cameron, J. R., Niroomand-rad, A. & Karam, P. A., 2002. Very high background radiation areas of Ramsar, Iran: preliminary biological studies. *Health Physics*, 82, 87-93.

---

1. It is also worth remembering that this variability is being found to be greater than was envisaged at the time of the deliberations that underpin current radiological protection principles.

[4] Hedin, A., 1997. *Spent nuclear fuel – how dangerous is it?* Swedish Nuclear Fuel and Waste Management Company (SKB), Stockholm, Technical Report TR-97-13, 60pps.

[5] ICRP, 1998. Radiation Protection Recommendations as Applied to the disposal of Long-Lived Solid Radioactive Waste. Publication 81. Annals of the ICRP, 28, No.4. Pergamon Press.

[6] INRA, 2002. Europeans and Radioactive Waste. Eurobarometer 56.2. Prepared for European Commission DG Energy and Transport. EC, DG Press and Communication, Brussels. 48 pps and Annex.

[7] NAPA, 1997. Deciding for the future: balancing risks, costs and benefits fairly across generations. Report for the USDOE by a panel of the US National Academy of Public Administration, Washington DC, USA. 49 pps.

1 000 YEARS OF SAFETY
TIMESCALES AND RADIATION PROTECTION REGULATIONS

Mikael Jensen

Swedish Radiation Protection Authority, SSI

1.   Background

A repository will exist forever in some form, and can therefore be assessed in many timescales with focus depending on the assessor's personal interest and role in the societal structure.

In a long enough timescale, any component in the performance assessment will defy a predictive description, although the point in time where science – gradually – breaks down as a meaningful tool for quantitative assessment may depend on the design, host rock, site and other circumstances. In any time perspective, the dose from a spent fuel repository may be higher than the dose from the excavated rock volume. However, this is a direct consequence of the decisions in connection with nuclear power production and the strategy to isolate the waste. It is therefore not meaningful to regulate dose or risk in the range of millions/billions of years. The regulations might still contain provisions for a qualitative description, and using natural analogues could provide valuable information in a "multiple lines of arguments" strategy to be presented in the licensing procedure.

The focus in many performance assessments lies on the period from closure and up to one or a few hundred thousand years. One driving force for the choice of this time period is the need to see unfolded all the components related to nuclide transport through the barrier system and the host rock. An understanding of those processes is necessary even if the regulation should focus on a shorter timescale.

In this work a third timescale is discussed, the first thousand years after closure. It usually receives very little attention, an unfair situation in the eyes of the author. Below an outline is given of SSI's regulations followed by a list of issues related to the 1 000 years post-closure period.

2.   SSI's regulations

2.1   Scope

SSI's regulations for disposal of radioactive waste use the term "final management" to include all back-end options of the fuel cycle, as well as final disposal options for other types of radioactive waste. The regulations therefore are general and the first paragraphs are not restricted to repositories. They cover all types of final management including transmutation and hypothetical disposal methods, e.g. sending waste into space etc.

2.2   *Holistic view*

SSI's regulations accept no discount for impacts outside Sweden's borders, and no discount for health effects to individuals or in the future.

They also require doses to be optimised, and they require collective dose to be reported for 1 000 years release integrated over 10 000 years. This part of the regulations is important because the general scope that includes a large range of management options in addition to final disposal.

2.3   *Protection of human health*

In the following, it is assumed that final management equals final disposal. For this solution, i.e. for a repository, the annual risk of harmful effects after closure must not exceed 10^{-6} for a representative individual in the group exposed to the greatest risk.

2.4   *The natural environment*

The regulations require that biodiversity and the sustainable use of biological resources are protected and that biological effects of ionising radiation in living environments and ecosystems concerned is described.

2.5   *Institutional issues*

The Swedish laws on radiation protection and nuclear activities do not contemplate a change of mind by a government in the future, including a distant future, reconsidering the basic purpose of the repository. The vice and virtues of any post-closure access or intrusion into the repository in the future are not addressed in the laws. The regulations therefore require that a repository be primarily designed with respect to its protective capability. However, it is required that the protective capability of the repository after a hypothetical intrusion is reported in the licence application, without any dose or risk limit attached to the description.

2.6   *Timescales*

The regulations require a quantitative description of the repository's impact on human health and the environment, for the first thousand years following repository closure. After that period, the assessment of the repository's protective capability shall be based on various possible sequences of the repository's development.

3.   The first thousand years

Most performance assessments indicate that releases will occur long after the first thousand years. Therefore, this period is usually not in focus. There are, however, several reasons for both regulators and operators to consider the first thousand years as a separate and important issue.

3.1   Jurisprudence

When engineered structures such as bridges and dams are discussed in society, e.g. in court proceedings, the time perspective is usually not more than 100 years. For radioactive waste repository, society expects engineers to perform at their very best and therefore to have the longest perspective reasonably possible.

In this context, the hundreds of years legal perspective can be seen as a motive for the thousand years period. It also follows that the extreme long timescales in performance assessment must be looked upon as very special cases, in any legal context.

3.2   Comparison with other radiation protection issues

The classical radiation protection time perspective is not, and should not, be in focus for a repository, but it may be worthwhile to make some general comparisons.

In comparison with other radiation protection issues, hundred years is long time. Radiation protection planning has the timescale of societal planning, from 10 to 10th of years. Some examples are:

- Chernobyl effects are often discussed over a time period of 100 years, and

- Carbon-14 release consequences from nuclear installations are usually given a sketchy treatment although the consequences could in principle build up during millennia.

An emphasis on a hundred years timescale would be too short for a performance assessment in the opinion of the author. It would not even cover the time span related to the phase "children and grandchildren". Perhaps the comparison should suggest a new look on long-term effects on Chernobyl and carbon-14 releases rather than shortening the assessment perspective for repositories.

3.3   Initial conditions

Some repositories, other than for spent fuel, are expected to experience a critical development during 1 000 years and are expected to have releases in this timescale. Releases are expected from the Swedish repositories SFR (for low and intermediate level waste) and SFL 3-5 (for long-lived wastes other than spent fuel) in 1 000 years. This therefore requires the whole chain of nuclide transport to be assessed in this time frame. However, it is an important confidence building issue for any repository that the operator knows and understands the initial conditions of the site.

It is true that the biosphere, subject to decisions in society, cannot be predicted, not even in the short term. But short-term scenario uncertainties are mainly related to man-made influences. With a "reference" society, the biosphere's evolution can be described or at least reasonably framed in a scale of thousand (up to about ten thousand) years. An item that can be reasonably well predicted is the process of land uplift by which sediment may be transferred into the bottom of a peat bog.

3.4   Other issues of confidence

The near future includes the lifetime of children and grandchildren. For this reason alone, it is a time of important concern for all. It may be argued that all periods are imbedded in each other so that long time periods include short periods, but the short-term focus may still appear to be lost in an assessment that only addresses the distant future. This is one of the main arguments for the emphasis of the thousand-year perspective in SSI's regulations.

3.5 *Hypothetical consequences of hypothetical releases*

Finally, the author would like to suggest a possible approach to address the short-term perspective, primarily as an issue of presentation. The implementer might ask: what would have to happen for a release to reach the biosphere in thousand years? The approach could be formulated in another way: list a number of features, events and processes (FEPs) for hypothetical short-term releases, and go through the list. In the selection of issues, the operator should consult the public to be sure that each and every concern is addressed. This must necessarily include low probability scenarios. The operator should not retract from such issues. Contrary to many people's doubts on the operator side, it is the author's experience that such a frank discussion rather will promote confidence than fear.

It might still be valuable to describe hypothetical consequences of hypothetical releases for the site. Such a process would demonstrate the operator's knowledge of the site and describe performance assessment techniques in a familiar setting.

FULFILMENT OF THE LONG-TERM SAFETY FUNCTIONS BY THE DIFFERENT BARRIERS DURING THE MAIN TIME FRAMES AFTER REPOSITORY CLOSURE

Peter De Preter and Philip Lalieux
ONDRAF-NIRAS, Belgium

In general terms the basic long-term safety functions of a disposal system (i.e. the engineered barrier system, including the waste forms and the host rock) are the functions that the system as a whole or its constituents must fulfil in order to assure an adequate level of long-term radiological safety.

The long-term safety functions of a disposal system constitute a generic and methodological tool that can be used in a double sense. In the first place these functions provide an *a priori* instrument for designing the system: the implementer must ensure that these safety functions are fulfilled by a series of robust system barriers and components. These functions can also be used as an *a posteriori* means to describe and assess in general terms the functioning of the system. In this way they are an important qualitative element to help to support the safety case and to identify further R&D priorities. By providing a general description of system functioning they are also a communication tool to stakeholders who are less familiar with the details of a safety case. Instead of limiting the description to a multi-barrier system, the safety functions enable to better explain how the different barriers contribute to one or more safety functions and by which processes this is performed. By doing so the system description moves from multi-barrier to multi-function.

The aim of this paper is to provide such a general description of the system functioning for the Belgian case of deep disposal of high-level waste (mainly spent fuel or vitrified waste from fuel reprocessing) in the Boom Clay, a poorly-indurated argillaceous formation. From the detailed safety and performance evaluations the main time frames after repository closure are identified. Each time frame relates to a period during which the successive safety functions play a key role. Also, in each time frame the radiological impact on the environment is distinctly different.

The long-term safety functions of the disposal system are "physical containment" (C), "delaying and spreading the release" (R), "limitation of access" (L); the long-term safety function of the environment of the disposal system is "dispersion & dilution"(D). The first two safety functions can each be subdivided in two sub-functions. For the "physical containment" function, which aims to prevent any release of activity from the waste matrix the two sub-functions are "water tightness" (C1) and "limit the water infiltration" (C2). The "water tightness" function must prevent any contact between infiltrating water and the waste matrix; the "limit the water infiltration" function must delay the moment that water can come into contact with the waste matrix without guaranteeing water-tightness.

The "delaying and spreading the release" function takes over when physical containment fails; it aims at slowing down radionuclide migration to the biosphere. By doing so, the system will mainly lower and spread the radionuclide releases into the biosphere in time. The two sub-functions are "resistance to leaching" (R1) and "diffusion & retention" (R2). The first sub-function spreads in time the radionuclide releases from the waste matrix; the second one delays the migration of the radionuclides to the biosphere and spreads in time the radionuclide releases into the biosphere. Slow diffusion-controlled migration, sorption on the solid (immobile) phase and precipitation/co-precipitation (low solubility limits) are the main underlying processes.

The safety function "limitation of access" aims to limit the likelihood and consequences of human intrusion directly into the closed repository. The depth and location of disposal away from natural resources, the resilience of the disposal system to intrusion and the preservation of the memory of disposal are all elements that contribute to this function.

Finally, radionuclide fluxes and concentrations are diluted and dispersed in the environment of the disposal system, i.e. in the aquifers and biosphere.

In a preliminary evaluation of the normal evolution of the disposal system in time, during repository operation and after repository closure, four characteristic periods or phases were identified, and which will be refined in future evaluations:

- the operational phase: this is the period between waste emplacement and repository closure, and can take several decades;

- the thermal phase: during this phase the heat generating waste significantly increases the temperature in and around the repository; its duration is approximately 300 years for vitrified waste and 2000 years for UO_2 spent fuel;

- the isolation phase: during this phase the radionuclide releases from the disposal system are negligible; in case of deep disposal of high level waste in the Boom Clay this phase is situated between 1 000 and 10 000 years after repository closure;

- the geological phase: here the repository enters the geological timescales (10 000 till million years after closure); the major impact, both due to non-retarded and retarded radionuclides is situated in this phase.

A central assumption in the normal evolution scenario and in the isolation phase is the stability of the Boom Clay barrier, meaning that radionuclide migration in this time frame does not change as no major changes occur in the Boom Clay.

For the normal evolution of the disposal system the contributions of the long-term safety functions in each of the four phases is given in Figure 1. The safety reserves in this figure are based on realistic estimates of container, overpack and waste matrix lifetimes as compared to the conservatively assessed (minimum) lifetimes.

A next step in the analysis is the identification of the different barriers and components of the disposal system and its environment that contribute to the long-term safety functions in each of the post-closure lifetime phases. A cross section of a disposal gallery for high-level vitrified waste is given in Figure 2. The result of this analysis is summarised in Table 1. This table enables to analyse and show. system redundancies on the functional level and to make visible the safety reserves, i.e. the contributions of system components to the safety functions that are not considered in the safety

Figure 1. **The four phases of the normal evolution of the disposal system (high level waste) and the corresponding long-term safety functions**

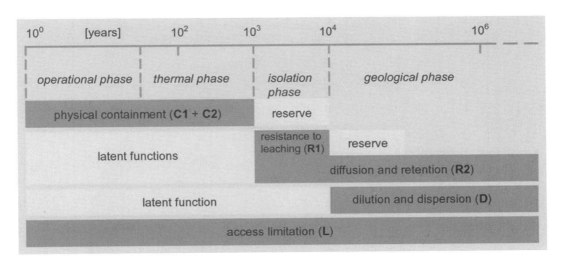

evaluations. This can be illustrated for the container or overpack barrier in the isolation phase. After degradation by corrosion the "physical containment" function is no longer assured, and the "delaying and spreading the release" function takes over. The corrosion products of the container or overpack will strongly sorb most of the radionuclides that are leached from the waste matrix; they contribute in this way to the "diffusion and retention" subfunction. As this process is not taken into account in the safety assessments it can be considered as a safety reserve.

Finally, this analysis also provides elements to prioritise future R&D work. For example the possible fragility of the system on the level of the physical containment safety function is an element that deserves a special attention.

On the other hand the "delaying and spreading the release" and "dilution and dispersion" functions are latent functions during the physical containment, because even in case of a premature failure of the physical containment the latent functions will perform partially or totally. These latent functions have to be evaluated in the altered evolution scenario of a premature failure of the physical containment, in order to assess if this early failure could have unacceptable radiological consequences. The performance of the "delaying and spreading the release" function in the case of this premature failure of containment can be affected by the higher temperature in the disposal system, resulting in a more rapid degradation of the vitrified waste matrix and an accelerated migration in a thermal gradient. This analysis of the latent functions still has to be done.

This analysis has focused on the normal evolution scenarios for which the safety functions are defined in the first place. The purpose of the altered evolution scenarios in a safety assessment is to evaluate or test if important disturbances of the system or unexpected failures of system components can significantly increase the radiological consequences compared to the consequences of the normal evolution scenario.

Future developments of this approach will focus on a refinement of the considered time frames, on the supporting arguments for splitting up in different time frames, as well as on redundancy and safety reserve evaluations.

161

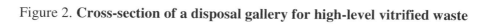

Figure 2. **Cross-section of a disposal gallery for high-level vitrified waste**

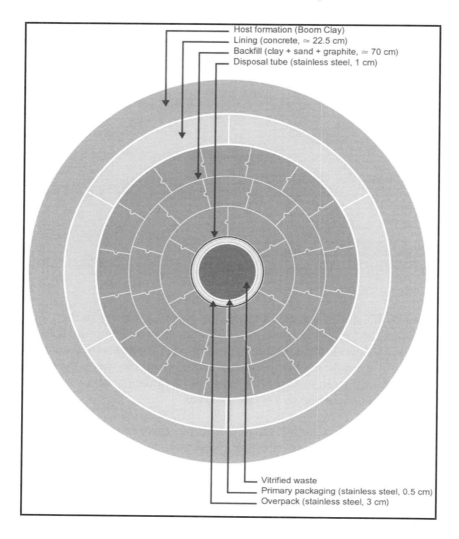

Table 1. Components of the disposal system and its environment that contribute to the long-term safety functions during the three main phases after repository closure. The functions that are considered in the safety evaluations are given in bold

Component	Thermal phase (1000 y)	Isolation phase (< 10 000 y)	Geological phase (> 10 000 y)
Glass matrix	–	**R1**	R1
UO$_2$ matrix	–	**R1**	**R1**
Vitrified waste canister	C1	R2	–
Container (spent fuel) or overpack (vitrified waste)	**C1**	R2	–
Disposal tube	C1	R2	–
Backfill in disposal galleries	C2	R2	R2
Seals	C2	**R2**	**R2**
Gallery lining	–	–	–
Shaft lining	–	–	–
Clay host rock	C2	**R2**	**R2**
Aquifers	–	**D**, R2	**D**, R2
Biosphere	–	**D**	**D**

TREATMENT OF BARRIER EVOLUTION: THE SKB PERSPECTIVE

Allan Hedin

Swedish Nuclear Fuel and Waste Management Co., SKB, Sweden

1.   Introduction

This paper serves as a point of departure for the discussions to be held within the Working Group of Technical Topic B entitled "Barriers and System Performances within a safety case: Their functioning and Evolution with Time". The paper gives the SKB perspective of the issues to be discussed within the Working Group for this Topic, as they have been described in the Workshop Programme.

The preliminary Workshop Programme presents the following issues to be discussed by the Working Group for Topic B:

- What is the role of each barrier as a function of time or in the different time frames? What is its contribution to the overall system performance or safety as a function of time?

- Which are the main uncertainties on the performance of barriers in the timescales? To what extent should we enhance the robustness of barriers because of the uncertainties of some component behaviour with time?

- What is the requested or required performance versus the expected or realistic or conservative behaviour with time? How are these safety margins used as arguments in a safety case?

- What is the issue associated with the geosphere stability for different geological systems?

- How is barriers and system performances as a function of time evaluated (and presented and communicated) in a safety case?

- What kind of measures are used for siting, designing and optimising robust barriers corresponding to situations that can vary with time? Are human actions considered to be relevant?

The following is a brief and preliminary account of SKB's view of these issues. The time frames for the presentation have been directly adopted from a Workshop presentation by De Preter and Lalieux entitled "Fulfilment of the long-term safety functions by the different barriers during the main time frames after repository closure". These authors divide the time after waste deposition into four different phases:

- the operational phase between emplacement and repository closure, extending over several decades

- the thermal phase (~1 000 yrs) during which heat generated by the waste increases the temperature of the host rock

- the isolation phase (~10 000 yrs) during which releases from the disposal system are negligible and

- the geologic phase (>10 000 yrs) during which the isolating capacity of the engineered barriers is no longer ensured

The KBS 3 concept with spent fuel in copper canisters surrounded by bentonite and deposited in granitic rock is in many respects different from the concept discussed by De Preter and Lalieux with both vitrified waste and spent fuel in steel canisters deposited in Boom clay. Nevertheless, in order to find a common ground for the discussions of timescales, a subdivision into the same phases will be used. The main difference lies in the fact that the KBS 3 system with its copper canisters is designed to isolate the waste over times extending into the geologic phase as defined above. Therefore, these two phases are treated together below. The operational phase will not be treated since this is outside the scope of the Workshop.

When discussing the different time frames, it is also important to bear in mind the evolution of the hazard of the waste, which decreases significantly over the time periods of interest.

2. Role of barriers

The roles or, rather, the safety functions in the three time phases of the different barriers or system components in the KBS 3 concept are presented in Table 1. In SKB's recent safety assessment, SR 97 [1], it is demonstrated that in the reference case, the copper canisters are expected to keep their isolating capacity intact throughout the time phases discussed above. The role of the other barriers or system components will in the reference case thus be to provide suitable conditions for the canister. SR 97 also treated a scenario with initially defective canisters due to e g imperfect sealing. Here, additional safety functions of the different barriers become important, most notably the retarding function of the buffer and the host rock. The safety functions for this scenario are, for the isolation and geologic phases, presented in a separate column in Table 1.

Table 1. **The safety functions in the KBS 3 concept assuming intact or defective canisters**

	Thermal phase 0 – 1 000 years	Isolation and Geologic phases; $0 – 10^6$ years	
		Intact canisters	Defective canisters
Fuel	As isolation phase	None	Confine matrix embedded nuclides
Canister	As isolation phase	Isolate	Limit release rate from canister interior

Table 1. **The safety functions in the KBS 3 concept assuming intact or defective canisters** (cont'd)

Buffer	As isolation phase + Conduct heat	Protect canister (keep canister in position, prevent advective transport, exclude microbes)	*Retard* *Conduct gas (H$_2$)* *Filter fuel colloids*
Backfill	As isolation phase	Confine buffer	*Limit advection in tunnels*
Geosphere	As isolation phase	Provide stable chemical and mechanical environment	*Retard*
Biosphere	As isolation phase	None	*(Dilute)*

3. Uncertainties

The main uncertainties concerning the barrier performances for the three time phases are presented in Table 2. Additional important uncertainties affecting barrier performance are related to the repository environment on a larger scale, most notably uncertainties related to future climate.

Table 2. **Uncertainties KBS 3**

	Thermal phase 0 – 1000 years	**Isolation and Geologic phases; 0-10^6 years**	
		Intact canisters	*Defective canisters*
Fuel			*Fuel dissolution rate*
Canister (copper shell and iron insert)		Quality of sealing Mechanical strength (isostatic load during glaciation, dynamic load at earthquakes)	*Evolution of initial defects*
Buffer		Chemical stability	*Gas transport properties*
Backfill			*Hydraulic conductivity, especially under saline conditions*
Geosphere		Chemical & mechanical behaviour during glaciation	*3D conductivity, transmissivity*
Biosphere	Societal structure, human behaviour	Societal structure, human behaviour	*Type of ecosystem, radionuclide turn-over, human behaviour*

4.   Performance evaluation

The performance of the different barriers over time are evaluated with a number of different methods. These include:

- thermodynamic arguments (stability of copper in Swedish deep ground waters);

- kinetic arguments (corrosion rate of iron, fuel dissolution rate);

- mass balance arguments (limited illitisation of bentonite, copper corrosion);

- natural analogues (long term stability of bentonite);

- long-term extrapolation of short term experiments/observations (corrosion processes, radioactive decay);

- complex modelling (groundwater flow, radionuclide transport, earth quakes).

Several of these may be used to evaluate the performance of a barrier for a particular time phase. Table 3 gives an overview of the most important methods currently used by SKB for a selection of key processes.

Table 3. **Performance evaluation KBS 3**

| | **Thermal phase 0-1000 years** | **Isolation and Geologic phases; 0-10^6 years** | |
| | | **Intact canisters** | |
			Defective canisters
Fuel			*Fuel dissolution: Experiments*
Canister (copper shell and iron insert)	Peak temperature: Model calculations	Copper corrosion: Thermodynamic & mass balance arguments Sealing: Testing of canister seals Mechanical stability during glaciation: Model calculations based on test results from manufactured canister inserts	*Iron corrosion: Experiments*
Buffer	(Saturation: Model calculations)	Chemical evolution and stability: TD model calculations, Natural analogues	*Gas transport: Experiments*
Backfill		See buffer	
Geosphere	(Resaturation: Model calculations)	Chemical & mechanical behaviour during glaciation: Model calculations	*Ground water flow: Hydro modelling based on 3D rock map from site investigations RN transport: 1D transport model with input from hydro modelling*
Biosphere			*RN transport modelling, Exposure pathway analysis*

5. Safety margins

For most safety functions in the KBS 3 concepts, the results of a realistic and also a pessimistic (conservative) evaluation of performance exceeds the required performance, and it is relevant to talk about a safety margin. This is described in the safety report wherever relevant, but there is no systematic way of using the safety margins in the final safety case in SKB's most recent performance assessment.

6. Geosphere stability

The meaning of "geosphere stability" is here taken to include both mechanical and chemical stability in relation to what is desirable for a KBS 3 type repository. Two main issues concerning stability of the Swedish granitic bedrock relevant for the KBS 3 concept can be identified. Both regard conditions related to glaciations, which are expected within the next several thousand years in Sweden:

- the maintenance of reducing condition where a particular issue is the possibility of intrusion of oxygenated groundwater during a glaciation and

- the effects of seismic events during a deglaciation.

7. Measures for siting, designing and optimising

Uncertainties concerning the long-term evolution of the repository system have influenced the siting and design of the repository in several ways.

A fundamental principle behind the design is the multi-barrier concept, through which the effects of a possible long-term deterioration of one barrier are mitigated by the presence of the other barriers.

Another principle underlying the design of the repository is the choice of materials which are stable over long time periods. This applies to copper as the canister material and bentonite as the buffer material. Copper is thermodynamically stable under conditions expected in deep groundwaters in Sweden and bentonite is a naturally occurring material which was formed of the order of 10^8 years ago and which has since then mostly been exposed to conditions similar to those expected in deep groundwaters.

The layout of the repository is chosen so that the distances to major fractures / faults, where future seismic events are expected to occur, are chosen so that the risk that these events will cause canister damage is kept low.

A further design consideration is that the distance between neighbouring canisters is chosen so that the maximum temperature on the canister surface is kept well below 100°C. This is to avoid boiling and accompanying enrichment of salts on the surface which could in turn cause long-term corrosion effects that are difficult to analyse.

Regarding siting, the repository will be located in a portion of the bedrock which does not contain ores of potential interest for future generations.

A more general principle behind the deep repository is to isolate the nuclear waste from man and the surface environment, thereby minimising the effects of uncertain future societal changes.

REFERENCES

[1]   Deep repository for spent nuclear fuel; SR 97 – Post-closure safety. Main Report Volumes I and II. Swedish Nuclear Fuel and Waste Management Co., Stockholm 1999.

PHENOMENOLOGY DEPENDENT TIMESCALES

Gerald Ouzounian
ANDRA, France

As required by the French act, Dec. 1991, construction projects for disposing of radioactive wastes have to be submitted to the Parliament by 2006. One of the most important points to allow for a decision at this time will be to gain confidence. The major difficulty in such a technical and societal project is to be able to carry out a demonstration of the safety over timescales which are out of the scope of any experiment. Among the arguments involved for the safety case are a series of simulations which objective is to assess the level of safety which can be reached, and its robustness to various internal defects (construction of the drifts, welding of canisters…) or external events (intrusion with deep boreholes, climate change, faulting,…).

Confidence in the simulations can be achieved if they are transparent, based on well understood processes. However, the complexity of the disposal system is such that temptation was great by the past to simplify the models, with a poor level of reporting on justifications, thus leading to what has been described as blackbox models. In the frame of the demonstration to be brought out for 2006, ANDRA has developed an approach consisting first to describe and analyse all the processes occurring over time and space in the repository. Once this type of information has been gathered in a structured way, then further analyses leading to abstractions, simplifications can be performed in order to facilitate simulations as required for the safety demonstration. The first stage of the approach has been called the phenomenological analysis of the repository situations (PARS). This work gives rise to a reference book in which our knowledge has been reported before being used for the safety demonstration. It also represent a reference for all technical and scientific knowledge based applications, such as digital modeling which is the basis for simulations, the repository design, the reversibility study, including the definition of a monitoring/observation program.

The requirements for the PARS are to fulfill completeness, thus satisfying safety-analysis requirements, and traceability. For this last point, the PARS is capable to evolve in a traceable way according to advances with regard to repository design and further phenomenological knowledge.

The principle of the method relies on a space and time segmentation of the repository evolution, giving rise to "situations", a situation being a phenomenological state of part of the repository or of its environment during a given period of time. For the Meuse / Haute-Marne site to which the PARS has been applied, the space and time segmentation was made possible through:

- The modular design of the repository.
- The layer structure of the geological medium.
- The identification of events triggering new phenomena during operations.
- Differences between the characteristic timescales of phenomena during all the lifetime of the repository.

One of the very important interests of this approach is the possibility to divide the modelling of a repository's phenomenological evolution, and to have simpler models to develop and then to handle for the simulation programme.

The PARS covers the evolution of the repository from the very beginning of the construction up to the appropriate times in relation with the radioactive decay of the waste (1 million years). The analysis of the repository in its environment is performed at a relevant scale in relation with its size and its environmental effects. The considered phenomena are thermics, hydraulics, mechanics, chemistry and radiology.

Input data for the first stage of the PARS are the preliminary designs of the repository, which integrate the state of acquired knowledge on the Meuse / Haute-Marne site and the waste inventory. The spectrum of preliminary designs investigated was selected in order to analyse the entire set of phenomena and factors likely to be involved in the repository evolution. Architectures have been selected for simplification purposes in understanding and modelling phenomena:

- Single-level repository in the middle of the Callovo-Oxfordian clay formation.

- Separation of disposal areas for B waste C vitrified waste, UOX spent fuel and MOX spent fuel , and fragmentation of respective areas into modules.

- An architecture consisting of a nest of "dead-end" cells to limit hydraulic connections.

- Maximum-temperature criteria determining the disposal density of exothermic waste (number of packages/cells, distances between cells).

Spatial fractioning of the repository system has been developed according to the main repository components, surface facilities, shafts, access drifts, 4 disposal areas: B waste, C waste and spent fuels, geological medium and the surface environment. Time fractioning of the repository evolution has been done according to an operational phase and a post-closure phase.

For each situation, the analysis consists in describing the phenomena (THMCR), their couplings and sequences, and identifying the potential radionuclide releases by the waste packages and defining their transfer pathways. Finally, an inventory of existing numerical models applicable to the situation has been proposed, as a first input to modelling and simulation.

It is proposed to introduce the first set of the 83 PARS files, and to illustrate it with a few situations as typical examples, with very explicit views. Conclusions of this first experience are then drawn in 2 ways:

- Management of pending indeterminations on the design and of uncertainties on phenomenological understanding of the situations as well as on data.

- Organisation of the modelling and simulation programme for the 2006 filing and reporting process.

LONG SCALE ASTRONOMICAL VARIATIONS IN OUR SOLAR SYSTEM: CONSEQUENCES FOR FUTURE ICE AGES

S. Edvardsson and K.G. Karlsson
Department on Physics, Mid Sweden University, Sweden

1.   Introduction

It is well known that the Earth has gone through a number of glaciations during the last few million years. It is generally agreed that changes in the Earth's orbital parameters and obliquity play an important role in climate forcing, an idea originally due to Milutin Milankovitch. The climatic history of the Earth shows clear evidence of the precession periods (about 19 000 and 23 000 years) and changes in the obliquity (period about 41 000 years). The main period in glacial data (around 100 000 years) is, however, hard to explain in terms of orbital variations.

During the last decade or so a large number of simulations of the planetary system have been reported, many of which have been concerned with the influence of variations in the Earth's orbit on climate (Berger and Loutre 1991, 1992, 1997; Berger *et.al.* 1993; Quinn et. al. 1991). In former studies comparisons have been made between the astronomical parameters and the ice volume of the Earth. Generally speaking, the match between astronomical forcing and ice data has been rather poor. In this study we instead compare the parameters with *changes* in ice volume, which yields a much better agreement between the astronomical and the geological data.

A number of climate models have been constructed to reproduce past climate history, and also to make predictions about the future (Imbrie & Imbrie 1980, Kukla *et al.* 1981, Gallée et al 1991, 1992, Berger *et al.* 1996). These are purely empirical models designed to match certain periods of past climate. Typically they also involve a substantial number of free parameters.

2.   Our model

In our model we have used several numerical methods. For a detailed description, see Edvardsson *et al.* (2002). The model is accurate because:

- Interactions from the Sun, the Moon and all planets are accounted for at every time step. To our knowledge, no other simulation has integrated the Moon orbit for such an extended time period.

- We have used state of the art numerical integration techniques. Several time steps have been used in order to verify complete convergence.

- We have derived an exact numerical treatment of the instantaneous spin axis. This property is computed at every time step, meaning that the model yields all oscillations from hours up to millions of years.

- When computing the total torque acting on the Earth, all other bodies in the system are taken into account at every time step.

- We have verified that general relativity, tides, and solar mass loss are negligible for times up to 1 Myr.

- The model has been tested to be reasonably insensitive to variations in solar mass, Earth mass, Moon mass, moments of inertia, and spin angular velocity.

3.   Climate considerations

The usual way for computing astronomical influence on climate is to use the eccentricity (e), the obliquity (θ), and the longitude of the perihelion (ω) to compute the climatic precession parameter $e\sin\omega$ and then to compute the insolation for various latitudes and time intervals. Since our model gives all the relevant information at each integration step we have adopted a slightly different method. At each step the direction of Earth's rotation axis is checked, and the time when it is directed towards the sun is defined as the summer solstice. For the calculations in this paper we are only interested in the summer solstice of the northern hemisphere. When the time of summer solstice is identified the corresponding values of obliquity (ε) and sun-planet distance (r) are printed to a file. Then the quantity $P = \sin\varepsilon / r^2$ is used as a measure of northern hemisphere summer insolation power. Following the argument by Edvardsson *et al.* (2002), the solar radiation power should be compared with the differentiated ice volume (changes). The ice volume data were obtained from the SPECMAP files (Imbrie et.al 1990).

As was mentioned above, empirical climate models have been developed to reproduce climate history and also to make predictions about the future. These phenomenological models involve adjustments of a number of free parameters. They are adjusted to certain periods of past climate. Usually the models show varying time lags between forcing and response and in the more sophisticated models various feedback mechanisms cause complications. Three models deserve mentioning:

- The ACLIN index (Kukla *et al.* 1981) is designed to predict the major climate changes in the late and middle Pleistocene and in the near future. It gives a fair reproduction of mild events, whereas cold events are not reproduced so well.

- Imbrie & Imbrie (1980) have developed a differential model that attempts to catch the essential features of the geological record for the last 150 000 years. The model involves the adjustment of four free parameters, and it simulates the last 250 000 years reasonably well. However, the last melting is not well reproduced, and there is a varying time lag between forcing and response.

- A sophisticated climate model based on astronomical climate forcing is the LLN model (Gallée et al 1991, 1992, Berger *et al.* 1996). The model simulates the northern hemisphere climate. Insolation variations force the model, but it also involves precipitation, evaporation, vertical heat fluxes, surface albedo, and oceanic heat transport. The extension of three continental ice sheets (Greenland, North America and Eurasia) are calculated separately. The model reproduces the Weichselian cycle well, but it must be noted that it involves adjustment of a large number of free parameters. Moreover, the sensitivity of simulated ice volume to CO_2 is not constant in time.

We have chosen not to construct a special climate model. Instead we perform a simulation into the future and directly integrate the quantity P as defined above to get an idea of the times and the amplitudes of future glaciations. Possible feedback mechanisms are discussed separately.

4.   Results and discussion

Figure 1 shows the variations of the Earth's obliquity ε, i.e. the angle between the Earth's spin axis and the instantaneous ecliptic, for the last 2 million years. The main period is around 41 000 years, and the maximum amplitude is around ± 1.3 degrees. We have compared our results with those of Berger & Loutre (1991), and with those of Laskar *et al.* (1993b). Up to 1 Myr the agreement between the different models is quite good, but differences increase considerably for times earlier than 1 Myr before present. A closer analysis reveals that the differences are mainly caused by a phase shift between different sets of results, while the amplitudes remain quite similar.

The bottom part of Figure 1 shows the variations of obliquity with and without the Moon. As has been pointed out by others (e.g. Laskar et.al. 1993a) the Moon is very important in stabilising the Earth. Without the Moon, the amplitude shows a tenfold increase compared to the case where the Moon is present. Without the Moon, however, the obliquity variations are also much slower.

The lower part of Figure 2 shows the ice volume as a function of time for the last 782 000 years with a resolution of 1 000 years. The upper part of the figure shows a comparison between the solar radiation P as defined above and the differentiated ice volume. Positions of peaks in the two curves are seen to coincide almost perfectly over the entire time interval. Not only the positions of the peaks in the two curves, but also the amplitudes, are quite nicely reproduced. Some deviations are observed at times of rapid melting at 10, 130, 340, 430, 620 and 740 kyr before present. It can be noted that the earlier problem of understanding the 100 000 year period in the ice volume data is not present in this comparison. Neither are there any time lag problems of 4-5 kyr between P and ice volume change, cf. Berger *et al.* (1993).

Figure 1. **(Top) Obliquity of the Earth over the last 2 Myr. The curve centred at 22 degrees shows the difference between the results of Berger & Loutre (1991) and ours while the curve centred at 21 degrees shows the difference between the relativistic results of Laskar *et al.* (1993b) and ours. (Bottom) Obliquity of the Earth with and without the Moon**

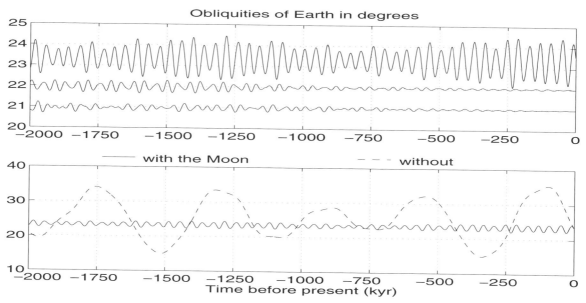

Figure 2. **(Top) Mean summer solar radiation power (insolation) and differentiated ice volume, <P>~dV/dt. (Bottom) Ice volume from Imbrie *et al.* (1990)**

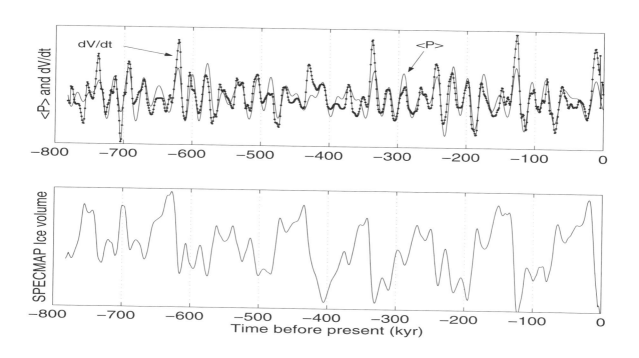

Figure 3 shows the integration of the solar radiation power P directly compared with the SPECMAP ice volume. A projection into the future is also provided. It is seen that the general oscillations and features of the ice volume curve are reproduced quite well. However, the deviations are sometimes substantial. In particular, the calculated ice growth after a major deglaciaton is exaggerated (e.g. at 230 and 120 kyr before present). The calculated meltings during the stadials at 160 and 30 kyr before present are exaggerated as well. These problems are of course due to the fact that our calculated radiation power in Figure 2 is not in perfect agreement with the differentiated ice volume. The reasons for the mismatches in Figure 2 are due to feedback mechanisms (greenhouse gases, changes in albedo etc.) during rapid deglaciations, volcanic activities and possibly external mechanisms such as changes in cloud cover due to cosmic radiation (Svensmark and Friis-Christensen 1997).

During interglacials large areas are covered by forests with a very low albedo probably making the Earth less sensitive to a decrease in solar radiation leading to the relatively slow growth of ice volume (Figure 2). In the end of a glacial when melting has been going on for a while, absorption of heat becomes more efficient due to a reduction of the ice cover and an increase in water vapour causing melting to accelerate. Moreover, the atmospheric CO_2 concentration shows a strong anti-correlation with ice volume, mainly because of changes in solubility of CO_2 with water temperature. When there is much ice, i.e. low temperature, there is much less CO_2 in the atmosphere. As temperature slowly rises during initial melting, CO_2 is released from the sea and the melted ice, causing a further increase in the temperature. We believe that these feedback effects largely explain the deviations during glacial terminations at 10, 130, 340, 430, 620 and 740 kyr before present observed in our Figure 2 (top).

Despite the above uncertainties, Figure 3 shows that there is a reasonable agreement between ice volume and calculations. It is therefore likely that there will be some ice growth peaking at around 22 000 years in the future, but that the magnitude of the peak might be exaggerated in Figure 3. On the other hand it seems highly likely that peaks at about 60 000 and especially at 105 000 years into the future will correspond to ice volumes close to those that prevailed during the last ice age.

Figure 3. **Integrated radiation power compared with ice volume and predictions for the future**

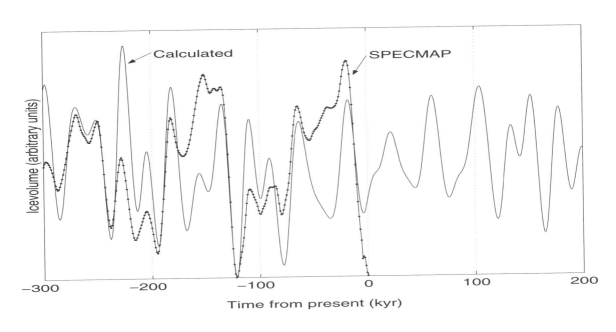

177

5. Conclusions

We have developed a numerical model of the solar system that seems to give reliable results for long integration periods. A straightforward method to compute summer solar radiation power has interesting applications for climatic studies. We find very good agreement between radiation power and *changes* in the ice volume of the Earth. Both build-up and melting of ice is found to agree well with the predictions inferred from the model. Some of the deviations can be understood in terms of feedback mechanisms. We conclude that the Earth will probably have considerable ice volume maxima at 60 and especially at 105 kyr after present. The computations also indicate a smaller maximum at 22 kyr after present. While the magnitudes of the maxima remain somewhat uncertain, we conclude that the timing of these events is quite certain.

REFERENCES

[1]   Berger, A., Loutre, M.F., 1991, Insolation values for the climate of the last 10 million years, Quaternary Science Reviews **10**, 297-317.

[2]   Berger, A., Loutre, M.F., 1992, Astronomical solutions for paleoclimate studies over the last 3 million years, Earth and Planetary Science Letters **111**, 369-382.

[3]   Berger, A., Loutre, M.F., 1997, Long-term variations in insolation and their effects on climate, the LLN experiments, Surveys in Geophysics **18**, 147-161.

[4]   Berger, A., Loutre, M.F., Tricot, C., 1993, Insolation and Earth's Orbital Periods, Journal of Geophysical Research **98**, No D.6, 10 341-10 362.

[5]   Berger, A., Loutre, M.F., Gallée, H., 1996, Sensitivity of the LLN 2-D climate model to the astronomical and CO_2 forcings (from 200 kyr BP to 130 kyr AP). Scientific Report 1996/1. Institut d'Astronomie et de Géophysique Georges Lemaître, Universite Catholique de Louvain.

[6]   Edvardsson, S., Karlsson, K.G., Engholm, M., 2002, Accurate spin axes and solar system dynamics: Climatic variations for the Earth and Mars, Astron. Astrophys. **384**, 689-701.

[7]   Gallée, H., van Ypersele, J.P., Fichefet, T., Tricot, C., Berger, A., 1991, Simulation of the Last Glacial Cycle by a Coupled, Sectorially Averaged Climate-Ice Sheet Model 1. The Climate Model, Journal of Geophysical Research, **96**, NO D7, 13 139-13 161.

[8]   Gallée, H., van Ypersele, J.P., Fichefet, T., Tricot, C., Berger, A., 1992, Simulation of the Last Glacial Cycle by a Coupled, Sectorially Averaged Climate-Ice Sheet Model 2. Response to Insolation and CO_2 Variations, Journal of Geophysical Research, **97**, NO D14, 15 713-15 740.

[9]   Imbrie, J., Imbrie, J.Z., 1980, Modelling the Climate Response to Orbital Variations, Science **207**, 943-953.

[10] Imbrie, J., McIntyre, A., Mix, A.C., 1990, Oceanic Response to Orbital Forcing in the Late Quaternary: Observational and Experimental Strategies in Climate and Geosciences, A Challenge for Science and Society in the 21st Century, Berger, A., Schneider, S.H., Duplessy, J.-C., eds., D. Reidel Publishing Company. Data from SPECMAP Archive #1. IGBP PAGES/World Data Center-A for Paleoclimatology Data Contribution Series # 90-001. NOAA/NGDC Paleoclimatology Program, Boulder CO, USA.

[11] Kukla, G., Berger, A., Lotti, R., Brown, J.P., 1981, Orbital signatures of interglacials, Nature **290**, 295-300.

[12] Laskar, J., Joutel, F., Robutel, P., 1993a, Stabilization of the Earth's obliquity by the Moon, Nature **361**, 615-617.

[13] Laskar, J., Joutel, F., Boudin, F., 1993b, Orbital, precessional, and insolation quantities for the earth from –20Myr to +10Myr, Astron. Astrophys. **270**, 522-533.

[14] Quinn, T.R., Tremaine, S., Duncan, M., 1991, A Three Million Year Integration of the Earth's Orbit, Astron.J. **101**, 2287-2305.

[15] Svensmark, H., Friis-Christensen E., 1997, Variation of cosmic ray flux and global cloud coverage – A missing link in solar-climate relationships, J. Atmos Sol.-Terr. Phys. **59**, 1225-1232.

THE SPIN PROJECT:
SAFETY AND PERFORMANCE INDICATORS IN DIFFERENT TIME FRAMES

Richard Storck and Dirk-A. Becker

Gesellschaft für Anlagen – und Reaktorsicherheit (GRS) mbH, Germany

1. Introduction

Safety and performance indicators have been under discussion for many years in several countries and international organisations. If those indicators refer to the long term safety of the total disposal system, they are often called safety indicators. If they refer to the performance of subsystems or the total system from a more technical point of view, they are sometimes called performance indicators. The need for indicators other than dose rates derives e.g. from the long time frames involved in safety assessments of waste disposal systems and the increasing uncertainty in dose rate calculations over time due to uncertainty in evolution of the surface environment and of behaviour of man.

Before introducing additional indicators into a safety case of a potential repository site, the applicability and usefulness of different indicators have to be investigated and evaluated. The systematic analysis and testing of safety and performance indicators for use in different time horizons after closure of the disposal facility is the task of the SPIN project. This is done by re-calculating four recent studies concerning repository projects in granite formations.

SPIN (Testing of Safety and Performance Indicators) is the name of a research project, funded by the European Commission within the fifth EURATOM framework programme "Nuclear Energy". It is carried out by eight organisations from seven European countries. These are:

- GRS, Germany;
- Enresa, Spain;
- Nagra, Switzerland;
- Colenco, Switzerland;
- NRG, The Netherlands;
- NRI, Czech Republic;
- SCK·CEN, Belgium;
- VTT, Finland.

The work is co-ordinated by GRS. The project runs from September 2000 to August 2002. This paper gives a short overview of the work performed and the results achieved up to April 2002.

2.   Definitions

It has to be clearly defined what is meant when talking about 'safety indicators' and 'performance indicators'. Since definitions of these terms have already been given by others, e.g. IAEA [1], it is not intended to give concurring definitions here, but some clarifications with respect to the special purpose of SPIN. In this context, safety and performance indicators are magnitudes following from numerical performance assessment calculations.

A safety indicator of the considered type must:

- provide a statement on the safety of the whole system;
- provide an integrated measure describing the effects of the whole nuclide spectrum;
- be a calculable, time-dependent parameter;
- allow comparison with safety-related reference values.

A performance indicator of the considered type must:

- provide a statement on the performance of the whole system, a subsystem or a single barrier;
- provide a nuclide-specific or integral measure;
- be a calculable, time-dependent or absolute parameter;
- allow comparison between different options or with technical criteria.

3.   Procedure

The testing and assessment of indicators is performed in three phases:

- identification;
- calculation;
- comparison;

For identification of indicators two different approaches have been used. In the first approach the open literature has been studied and examined for proposals of indicators. These were supplemented by own proposals where it seemed sensible. The other approach to identifying performance indicators is a systematic one, following so-called safety functions, which describe the basic functionalities of a deep underground repository. Safety functions are:

- physical confinement;
- retardation / slow release;
- dispersion / dilution.

In the calculation phase, the identified indicators have been calculated for four different studies of repositories in granite. The studies differ in various details and parameters, like waste type, type of containers, nuclide inventory, source term, thickness of buffer, geosphere properties, and treatment of biosphere. The nuclide data, however, have been harmonised in order to get comparable results. To allow calculation of the indicators, the available assessment and post-processing tools have been modified where necessary, then the studies have been re-calculated. The four studies are:

- ENRESA-2000 (Spain);
- GRS-SPA (Germany);
- Kristallin-I (Switzerland);
- TILA-99 (Finland).

For comparison purpose, all results have been presented in a common style, and, where necessary, in common figures. In a first step, the results have been compared with one another to assess their suitability for showing the safety of the system, or specific aspects concerning the performance of barriers. In the second step reference values have been identified which can be used for safety indicator results to compare with. These reference values have been taken from different sources. There are administrative nuclide flux constraints given by the Finish regulation authority STUK. Natural fluxes and concentrations are being determined in the Co-ordinated Research Programme (CRP) on Safety Indicators of IAEA, which has been used as a source of input for SPIN. Also other sources have been used, e.g. the national constraints for effective dose rates. This procedure shows whether it is or is not possible to give safety-related reference values for the safety indicators.

4. Performance indicators

Most performance indicators identified for testing are expressed in terms of either activity or radiotoxicity. Radiotoxicity is calculated by multiplying each nuclide's activity by its specific ingestion dose coefficient (IDC) provided by ICRP [2]. If nuclide-specific investigations are done, activity should be preferred, because it is the most direct physical measure for nuclide inventory. If, however, integral statements on the entire nuclide spectrum are to be derived, a proper weighting scheme is necessary to take account of different radiological effectiveness.

To allow comparison of performance indicator results for different studies, some generic subsystems have been defined which can be found in each of the considered repository systems. These subsystems are called compartments. The following compartments have been identified:

- the waste form, including only the waste itself;
- the waste package water and precipitate;
- the waste package, including all above and the container material;
- the bentonite buffer;
- the near field, including all above;
- the geosphere;
- the biosphere.

Most performance indicators identified for testing can be classified in four types. Each one has been tested in its activity and radiotoxicity forms:

- inventory in compartments [Bq or Sv];
- flux from compartments [Bq/y or Sv/y];
- time-integrated flux from compartments [Bq or Sv];
- inventory outside compartments (outside the outer compartment boundary) [Bq or Sv].

Additionally, the following indicators have been calculated:

- concentrations in compartment water (only for specific compartments) [Bq/m³];
- nuclide-specific transport times through compartments [y].

Figure 1 shows graphically the different compartments and table 1 their use for different types of performance indicators. Each of the indicators has its specific properties and advantages and should be used preferably for specific investigations. A detailed assessment, however, is not intended to be given within this paper.

Figure 1. **Graphical presentation of compartments and performance indicators**

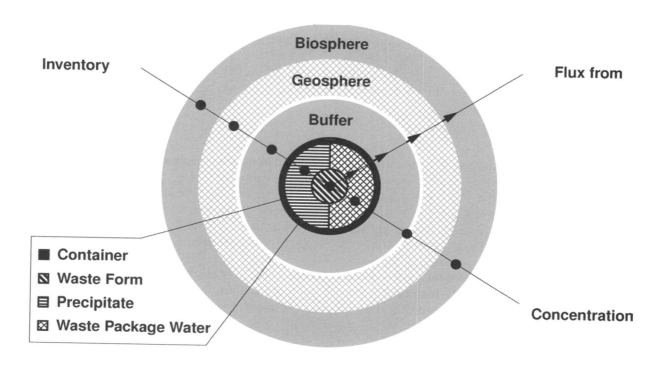

Table 1. **Compartments used for different types of indicators**

Compartment	Inventory in	Flux from/ Inventory outside	Concentration in	Transport time
Waste form	waste form	waste form		
Container	precipitate	waste package	waste package	
Buffer	buffer	near field		buffer
Damaged zone			EDZ water	
Geosphere	geosphere	geosphere		geosphere
Biosphere	biosphere		biosphere water	

5.   Safety indicators

While performance indicators are used to show in detail how the system works, safety indicators should demonstrate the overall safety of the system. Therefore, they are neither related to compartments, nor should they be used for nuclide-specific investigations. A proper nuclide-weighting scheme is needed for each safety indicator. This can be the set of IDCs or any other set of nuclide-specific weighting factors, e.g. a set of constraint values.

The following safety indicators have been identified for testing:

- Effective dose rate [Sv/y];
- Radiotoxicity outside geosphere [Sv];
- Time-integrated radiotoxicity flux from geosphere [Sv];
- Radiotoxicity concentration in biosphere water [Sv/m³];
- Radiotoxicity flux from geosphere [Sv/y];
- Relative activity concentration in biosphere water [-];
- Relative activity flux from geosphere [-].

Each of these indicators is calculated with a weighting scheme which takes account of the different hazard potentials of different nuclides, resulting in an overall sum which can be compared with a reference value. For radiotoxicity indicators this scheme is given by the IDCs. The last two indicators, however, need complete sets of constraint or reference values as weighting factors. Each nuclide's concentration or flux is divided by the relevant reference value, and the sum of all should not exceed 1. The relative activity fluxes have been calculated with constraint values provided by STUK, but these values are specifically made for the Finish situation and have low relevance for the other studies. Relative activity concentrations could not be calculated at all because a complete set of reference concentrations is not available.

The figures 2 through 5 show the calculated results of four of the indicators.

For assessment of the usefulness of safety indicators three groups of criteria have been identified:

Basic criteria used for identification of a safety indicator:

- It must be uniquely defined.
- It must be applicable in performance assessments.
- It must be safety related.

Evaluation criteria used for selection of indicators:

- Does it exclude biosphere pathways? The uncertainty of the pathways increases in time.
- Does it exclude dilution in aquifer water? This is an uncertain phenomenon.
- Does it exclude ingestion dose coefficients? They are made exclusively for human beings of today.
- Does it use a safety related weighting scheme? This is important for a statement on the safety.
- Are possible reference values safety-related and available? These are actually two different questions which should both be answered with 'yes'.

Final requirements used for final assessment:

- Is the indicator fairly easy to understand?
- Does it provide added value to other indicators?

Table 2 shows a schematic assessment of the safety indicators identified above, using the mentioned criteria. A + sign indicates that the criterion is fulfilled, a.- sign that it is not. A question mark indicates that the answer is not unique, e.g., for the relative activity concentration in biosphere water neither a weighting scheme nor a reference value could be identified, and therefore no statement on their safety relevance is possible. The basic criteria are omitted here because they have already been used in the identification process and are fulfilled by each of the indicators.

The different safety indicators seem to be of different suitability for different time frames. The effective dose rate is a very good indicator for the short term, but since the biosphere pathways have a limited validity, its usefulness decreases for longer time periods. The aquifer dilution will also vary with time, but is considered more stable. Indicators, which include the dilution but not the biosphere pathways, may be useful for medium timescales. If both uncertainties are excluded, the indicators are good for long timescales. The relative activity concentrations and fluxes could be good indicators for long or even very long time frames if appropriate weighting schemes and reference values were available. These statements are compiled in table 3 for the most important indicators.

Table 2. **Assessment of safety indicators**

Indicator	Eclusion of			available	Safety related	available	Safety related	Easy to understand	Added value
	Biosphere Pathways	Aquifer Dilution	Dose Coefficients	Weighting scheme		Reference value			
Effective dose rate	-	-	-	+	+	+	+	+	+
Radiotoxicity outside geosphere	+	+	-	+	+	+	-	-	?
Time-integrated radiotoxicity flux from geosphere	+	+	-	+	+	+	-	-	?
Radiotoxicity concentration in biosphere water	+	-	-	+	+	+	+	+	+
Radiotoxicity flux from geosphere	+	+	-	+	+	+	+	+	+
Relative activity concentration in biosphere water	+	-	+	-	?	-	?	+	?
Relative activity flux from geosphere	+	+	+	+	+	+	+	+	-

Table 3. **Time frames for safety indicators**

Indicator	Eclusion of			Time frames
	Biosphere Pathways	Aquifer Dilution	Dose Coefficients	
Effective dose rate	-	-	-	Short term
Radiotoxicity concentration in biosphere water	+	-	-	Medium term
Radiotoxicity flux from geosphere	+	+	-	Long term
Relative activity concentration in biosphere water	+	-	+	Long term*
Relative activity flux from geosphere	+	+	+	Very long term*

* Provided that safety-related weighting schemes and reference values are available.

For the usage of the identified safety indicators, the preliminary recommendations as shown in Table 4 can be given. The final assessment of all tested indicators, however, is still to be done.

Table 4. **Preliminary recommendations for use of safety indicators**

Indicator	Recommendation
Effective dose rate	Continued application for all time periods higher weighting at early time periods
Radiotoxicity outside geosphere	Not to be used, unless safety related reference values are identified
Time-integrated radiotoxicity flux from geosphere	
Radiotoxicity concentration in biosphere water	Preferred application for later time periods
Radiotoxicity flux from geosphere	Preferred application for very late time periods
Relative activity concentration in biosphere water	Not to be used, unless a new weighting scheme will be found
Relative activity flux from geosphere	Apply if you are asked to do so

Figures 2 through 5 show the results of the four safety indicators recommended for use, calculated for the four studies under consideration, each together with a reference value.

Figure 2. **Effective dose rate**

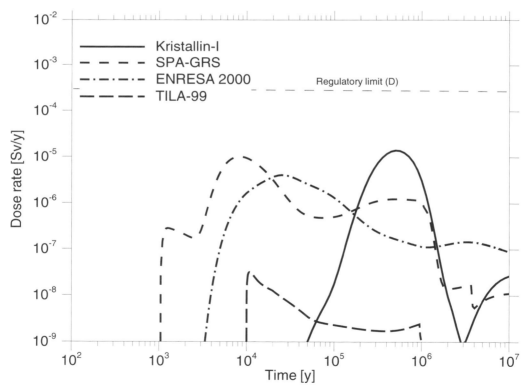

Figure 3. **Radiotoxicity concentration in biosphere water**

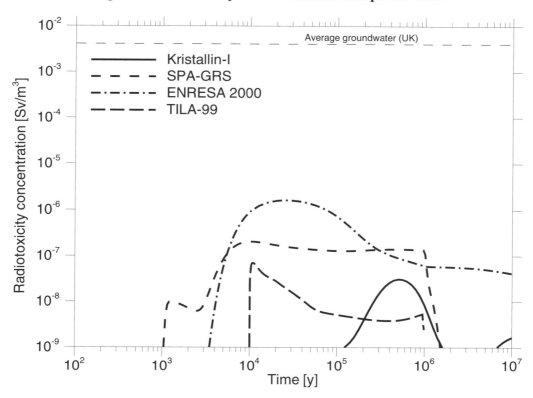

Figure 4. **Radiotoxicity flux from geosphere**

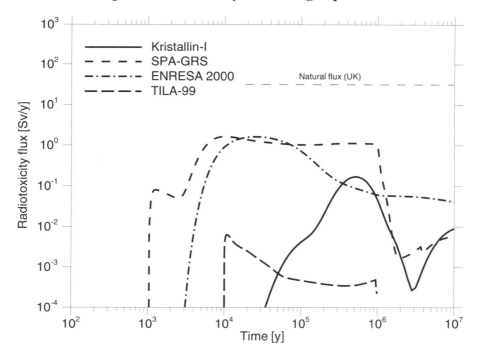

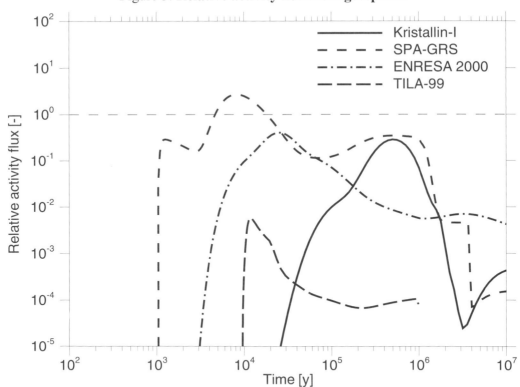

Figure 5. **Relative activity flux from geosphere**

REFERENCES

[1] Safety indicators, complementary to dose and risk, for the assessment of radioactive waste disposal: working material, Draft Report/Version 1; reproduced by IAEA, Division of Radiation and Waste Safety, Vienna 1999.

[2] International Commission on Radiological Protection: ICRP 72. Pergamon Press. Oxford, 1996.

IAEA ACTIVITIES RELATED TO SAFETY INDICATORS, TIME FRAMES AND REFERENCE SCENARIOS

B. Batandjieva, K. Hioki and **P. Metcalf**
Waste Safety Section, Division of Radiation and Waste Safety
International Atomic Energy Agency, IAEA, Austria

Abstract

The fundamental principles for the safe management of radioactive waste have been agreed internationally and form the basis for the Joint Convention on the Safety of Spent Fuel Management and on the Safety of Radioactive Waste Management that entered into force in June 2001. Protection of human health and the environment and safety of facilities (including radioactive waste disposal facilities) are widely recognised principles to be followed and demonstrated in post-closure safety assessment of waste repositories.

Dose and risk are at present internationally agreed safety criteria, used for judging the acceptability of such facilities. However, there have been a number of activities initiated and co-ordinated by the International Atomic Energy Agency (IAEA) which have provided an international forum for discussion and consensus building on the use safety indicators which are complementary to dose and risk. The Agency has been working on the definition of other safety indicators, such as flux, time, environmental concentration, etc.; the desired characteristics, and use of these indicators in different time frames. The IAEA has focused on safety indicators related to geological disposal, exploring their role in the development of a safety case, evaluating the advantages and disadvantages of using other safety indicators and how they complement the dose and risk indicators. The use of these indicators have been discussed also from regulatory perspective, mainly in terms of achieving reasonable assurance and confidence in safety assessments for waste repositories and decision making in the presence of uncertainty in the context of disposal of long-lived waste.

Considerable effort has also been expended by the Agency on the development and application of principles for defining critical groups and biospheres for deep geological repositories. One of the important and successful IAEA programmes in this field is the BIOsphere Modelling and Assessment (BIOMASS) project which was completed last year. The project proposed a methodology for the development of reference biospheres which is in line with suggestions of the International Commission for Radiological Protection 81 (ICRP). Its application is illustrated by way of three example biospheres. At present the Agency is continuing the work on development of reference scenarios for waste disposal in order to assist in the development of international agreement on the main principles and concepts for development and use of reference scenarios for use in the development of safety cases and decision making related to waste repositories.

1.   Introduction

The fundamental safety principles of radioactive waste management have been agreed internationally and are recognised as a basis for the Joint Convention on the Safety of Spent Fuel Management and on the Safety of Radioactive Waste Management that entered into force in June 2001. Protection of human health and the environment and the safety of facilities (including radioactive waste disposal facilities) are internationally established principles to be followed and demonstrated at all stages of radioactive waste disposal facility development, including the post-closure phase. Demonstration of the long-term safety of radioactive waste disposal facilities is required both in terms of formal regulatory authorisation programmes and for broader stakeholder acceptance. Because of the long-lived nature of some of the radioactive waste that has to be disposed of, it is necessary to demonstrate the safety of facilities for time period far into the future. In this regard there is a need to develop consensus on what those time periods should be. In addition, it is recognised that in very long time frames the criteria against which to judge the acceptability of facilities may not be the same as for shorter-term radiation protection considerations. Hence a considerable amount of work has been undertaken at the international level to consider what are appropriate indicators of safety in the longer term and over what time frames consideration should be given to demonstrating compliance with safety criteria.

2.   Safety indicators complementary to dose and risk

The key requirement for radioactive waste disposal is not to pose a significant hazard to human health or the environment now and in the future, even over long time periods in the future. Safety indicators are the characteristics or consequences of a disposal system by which hazards or harm can be measured and expressed, and may be evaluated in the course of making a safety case for a disposal system. The most widely used indicators in the context of the long-term safety of disposal systems are those of radiological dose and risk. The use of dose or risk as safety indicators has one main disadvantage, which is the associated uncertainty surrounding their estimation. Biosphere modelling is an integral part of dose and risk evaluation to humans, and uncertainties associated with biosphere modelling increase considerably with the time-scale under consideration, in particular with regard to human habits, but also with regard to the nature and distribution of other organisms. The engineered barriers of a disposal system and the geosphere, which will be characterised in the safety assessment process has in general, greater long term stability.

A particular disadvantage with the use of risk as a safety indicator is that risk values are not necessarily useful in communicating with the public when attempting to illustrate the safety of disposal facilities. Risk can be a difficult concept to explain to people and, for example, there is a tendency for much more attention to be paid to the consequences of an event than to the probability with which it is expected to occur.

Possible ways to circumvent the uncertainties inherent in biosphere modelling include the use of complementary safety indicators and the use of stylised approaches (e.g. one or more defined reference biosphere). Combinations of these approaches are also possible in the development of a safety case for a disposal facility. It should be pointed out that safety assessments making use of either one of the two possibilities outlined above are conceptually similar as they both reduce the biosphere-related uncertainties from the analysis. The choice of method will then depend on what is believed to be the most effective in presentation of the safety case. These considerations also imply that all safety assessment results, when they refer to a time sufficiently far in the future, cannot be considered as actual predictions of impacts, but are, instead, indicators of safety.

Complementary safety indicators may go some way to overcome the limitations of dose and risk discussed in the previous section. Their main potential advantage is that they are derived, in most cases, from calculations of radionuclide migration and distribution in the relatively stable medium of the geosphere, thereby eliminating from the assessment that part of the system (the biosphere) characterised by the most intractable uncertainty. There are, however, other potential advantages to using complementary safety indicators over and above the fact that they avoid the need to make assumptions about future biospheric and demographic conditions in estimating the values of the indicators themselves. The yardsticks against which complementary indicators are compared may also be based on observations of natural systems, which are generally characterised by long term stability. Complementary safety indicators could also alleviate some of the concerns associated with the application of radiation protection principles to the impacts of radioactive waste disposal in the remote future, such as uncertainty about the future validity of the assumed relationship between dose and detriment. Complementary indicators can provide for flexibility, diversity and transparency for a wide range of stakeholders (technical and non-technical). They provide the possibility for comparison with issues better understood by these stakeholders, for example, background radiation levels, fluxes and concentrations of naturally occurring radionuclides and other natural or man-made hazards.

Nonetheless, although complementary safety indicators offer a number of benefits, they also have some disadvantages, the most obvious of which is the lack of international consensus on how to apply them within a safety case. Other practical problems relate to the lack of information on natural radionuclide concentrations and fluxes, and their variability, for use in the development of yardsticks. This problem is, however, being addressed by ongoing projects to quantify more accurately natural radionuclide abundances and distributions

3.   Safety indicators and the safety case

The concept of a safety case for structuring and presenting the safety arguments for various nuclear and radiation technology related facilities have evolved in recent years. The safety case is seen to be made up of a series of arguments justifying the design and mode of operation of a facility, and of particular importance in respect of a waste disposal facility, its closure. The safety arguments are supported by quantitative assessment of the performance of various components of the facility and of the performance of the whole facility – the disposal system in respect of a radioactive waste disposal facility. The outcome of these quantitative assessments can be expressed in terms of annual radiation dose that may arise from normal operation or evolution of the facility, or the risk from disruptive events that can be anticipated but which are likely to be infrequent. Also, in the case of a waste disposal facility, assessments can be made of the possible consequences of human intrusion events. The safety case must also present the arguments for the level of confidence in the assessments, which can include arguments concerning the design logic, the choice of site, the choice and configuration of engineered barriers, the extent of multiple barriers and complementary safety functions and a consideration of the various uncertainties inherent in any assessment of this nature. The confidence arguments should also give consideration to the quality management programme applied to the siting, design, development, operation and closure of the facility.

The safety case can be used to present and justify the safety of the facility to a range of interested parties, but it is intended primarily to present the safety argument to the regulatory body with responsibility for authorisation of the facility, its operation and its closure. In this respect one important element that the safety case must address is compliance with regulatory requirements and criteria. International consensus over the nature and form of these requirements is presently being developed, but a consensus appears to be emerging that these should be a blend of quantitative criteria such as radiation dose and risk constraints, non-quantitative criteria such as a requirement to demonstrate system robustness and engender multiple safety functions and a requirement to develop a deep understanding of the disposal system and the phenomena influencing its performance. In addition, consensus appears to have developed that the safety case for a disposal facility, whilst, necessarily having to demonstrate the ability to respect radiation dose and risk constraints, should also contain multiple lines of reasoning regarding the safety of the facility and its robustness. In this respect safety indicators other than dose and risk are being brought into the formal safety case required by regulatory bodies.

In one of the earlier IAEA reports on safety indicators "Safety indicators in different time frames for the safety assessment of underground radioactive waste repositories" TECDOC-767 (1994), various potential indicators were identified and discussed in terms of their potential utility in different time frames. The indicators identified at that stage are still considered to be the more likely parameters of use namely; fluxes through the geosphere and biosphere, time frames for transport of mineral species through geological barriers or barriers engineered from natural materials, environmental concentrations of minerals and radio toxicity indicators related to natural radioactive materials such as mineral ores.

The same report concluded that for time frames up to 100 000 years, radiation dose and risk estimates were probably the most appropriate basis for regulatory consideration, but these could be supported by other indicators. For the period extending to 10^6 years the consideration was that dose and risk became less and less meaningful and that alternative indicators could play a more important role. Beyond 10^6 years it was concluded that the relevance or credibility of any assessment was questionable. In a later report "Regulatory decision making in the presence of uncertainty in the context of disposal of long lived radioactive waste" TECDOC-975 (1997) the concept of "reasonable assurance" of safety being demonstrated to the regulatory authority was emphasised, recognising the inevitable uncertainties attached with predictive assessments of performance and impact far into the future. The use of indicators was discussed in this report and their potential contribution to providing the necessary level of reasonable assurance of safety. The same indicators were recognised to be relevant but at the same time the difficulties in setting quantitative criteria for comparison with was recognised. It was also emphasised that the prime consideration is the limitation of radiological impact to humans and the environmental and that sight of this fact should not be lost.

In the most recent draft report on safety indicators "Safety indicators, complementary to dose and risk, for the assessment of radioactive waste disposal" (December 2001) the previous work has been continued to address the development of indicators and their possible application. Their complementary nature to dose and risk remains a constant consideration. Their contribution to enhancing confidence within broader stakeholder groups is also emphasised, but little change to their use in the safety case and regulatory decision making processes is suggested. The report also recognises that whilst the focus of use for safety indicators other than dose and risk was originally on geological disposal facilities, they could be of use in considering the safety of other disposal facilities including near surface facilities and mining and minerals proceeding waste disposal facilities.

The process of developing international safety standards for geological disposal facilities has been underway for around a year now and involved a Specialists Meeting, which took place in June 2001 and considered a number of issues including "Making the safety case – demonstrating compliance" and "Safety Indicators". With regard to the safely case, there was clear consensus that it should present multiple lines of reasoning and that safety indicators should be addressed together with their use. Nevertheless it was also concluded that it was not appropriate to set down explicit and quantitative regulatory criteria and that their use should be made in a defined context and for particular time frames.

4. Reference biospheres and scenarios

It has been recognised that the use of dose and risk constraints as indicators based on human health detriment from exposure to ionizing radiation can only be applied with any degree of certainty for a relatively short-period of time, in the region of hundreds of years. Predictions of future human behaviour patterns and the possible reliance on transfer of knowledge about current disposal activities has also been recognised to introduce significant uncertainties. Institutional control over waste disposal facilities is a safety measure that is more relevant for near surface repositories than geological disposal facilities and gives rise to problems in respect of uranium and mining/mill tailings because of the very long time periods, virtually indefinite, associated with the disposal of the associated long-lived radionuclides.

The impact of a waste repository could be evaluated on the basis of a conceptual (stylised) critical group with assumed habits and characteristics. As the humans who make up such critical groups are part of the biosphere, they cannot be separated, and assumptions made are interlinked and in this regard the characteristics of the biosphere should support and be consistent with the assumed critical group.

For long-time periods, where there are considerable uncertainties associated with the probability of occurrence of events, such as human intrusion, and the evaluation of their consequences, ICRP 81 recommends the use of stylised intrusion scenarios. A series of activities related to the development of stylised biospheres has been undertaken during the last years by a number of countries in projects such as BIOMOVS. The IAEA has continued the efforts in this field as part of the BIOMASS project that was completed last year. The BIOMASS project aimed to develop the concept of reference biospheres into a practical system, which could be applied in the long-term safety assessment of waste repositories. In the first place a methodology for development of reference biospheres has been internationally agreed that provides a systematic, logic and transparent staged approach to the development and justification of the biospheres to be considered in safety assessment. The methodology involves several main elements, which are presented in Figure 1.

Figure 1. **BIOMASS methodology**

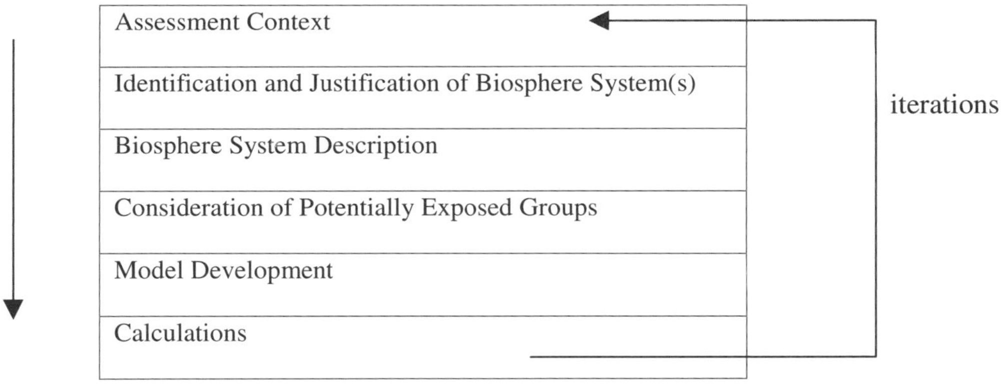

To develop a suitable biosphere for a waste disposal facility requires consideration of a number of issues such as the purpose of the safety assessment, general requirements, calculational endpoints, the waste inventory, geosphere-biosphere interface, time frames considered, assumptions about the society and levels of conservatism. All these aspects need to be addressed as a first step - in the defining the so-called *assessment context*. Based on the assessment context it is important *to identify and justify the biosphere* system through identification of the main components of the biosphere system (climate, topography, human activities, etc.); consideration of changes of the biosphere with time; the way of biosphere change (e.g. set of unchanging biospheres, time-related transfer from one biosphere to another). Once the biosphere has been identified it is necessary to provide an adequate and detailed description of the system(s), including identification of potentially exposed groups (critical groups). For each component of the system, lists of important features, events and processes are screened to define the ones that are important for long-tern safety assessment. Subsequently these features events and processes serve as a basis for the development of conceptual and mathematical models. In this respect, the use of a radionuclide transfer matrix has proved to be a very useful tool to describe the relevant conceptual models. The availability of information and data is very important in establishing a mathematical representation of the biosphere system – *mathematical modelling*. The availability of adequate data allows performance of *calculations* to estimate the radionuclide concentrations in different media (water, soil, etc.) and relevant doses, risk to the critical group or other endpoints considered.

Application of the BIOMASS programme has been demonstrated in the development of a set of reference biospheres related to an unchanging temperate climate and in a defined set of conditions:

a) drinking water well intruding into a contaminated aquifer;

b) agricultural irrigation well intruding into a contaminated aquifer;

c) natural discharge from a contaminated aquifer into a number of different habitats, including arable, pasture, semi-natural wetland and lake.

The programme has also involved consideration of using the methodology for development of reference biospheres with conditions that change with time, the examples being based on two existing sites – Harwell (United Kingdom) and Äspö (Sweden) for which information and data is available.

BIOMASS has demonstrated that the methodology developed by a broad group of experts made up of regulators, operators, and independent experts, has proved to be both practical and consistent with the recommendations of ICRP 81 on radiation protection and disposal of radioactive waste. The methodology provided a systematic approach for use in decision making, including choice of critical groups and addressing change of the biosphere with time. Recognizing the importance of input data, BIOMASS has also developed a data protocol, which provides a systematic approach to data selection and documentation of the assumptions made in performing the assessment. Although the methodology was mainly focused on human exposure as a result of radionuclide transport through groundwater, it is considered relevant to test the use of other endpoints. The outcome of the development of biosphere concepts with increasing complexity and including different ranges of habitats and groundwater interfaces with the biosphere serves as a basis for investigation of the acceptance of the internationally agreed biospheres as a reference to be used in decision making related to the safety of waste disposal facilities.

Making use of the outcomes from BIOMASS, the IAEA has also undertaken work on the development of reference scenarios for waste disposal facilities. The objective is to investigate the development of a set of straightforward scenarios based on international consensus that will assist communication and confidence building in different stakeholders and address the long-term uncertainties in safety assessment. It will at the same time provide a benchmark for comparison of waste disposal facilities. It is expected that these scenarios will enable preliminary evaluation of human impact, based on the concentrations of radionuclides in the geosphere. Four scenarios are being considered for further investigation – drinking water, borehole penetrating the waste vault, farming on waste contaminated soil, and housing (residence) on land contaminated with radioactive waste. The development of reference scenarios will continue and the results will be presented in a working document for broader discussion.

5.   Conclusions

Indicators, together with appropriate yardsticks, can be used to provide safety arguments in support of the acceptability of a disposal system, and the quality and reliability of its various components, including the suitability of a chosen site and design to provide long-term isolation. The use of safety indicators other than dose and risk complement, rather than supplant, the use of dose and risk; recognising the distinct advantages, in terms of avoiding sources of uncertainty, related to future human behaviour and for communicating to wide audiences. In using this approach in the development of a safety case for a waste disposal facility it is important to ensure that the suite of indicators selected are consistent with the waste and disposal concept, the assessment basis and the requirements and expectations of the potential audiences for the safety case. It is also important to use a variety of safety indicators in the safety case in a way that the limitations of one are balanced by the strengths of another.

The application of complementary safety indicators is expected to be of most value when evaluating the consequences of disposal facility evolution scenarios, in which radionuclides released from the near-field are transported to the accessible environment (biosphere) by natural processes (e.g. groundwater discharge). The approach is unlikely to be useful when evaluating the consequences of scenarios in which the geological barrier is bypassed, such as those involving human intrusion or natural disruptive events at the waste disposal facility system. Among the variety of potential indicators discussed, concentrations and fluxes in the geosphere and surface environment, and containment timescale appear to have the greatest potential as complementary indicators of overall system safety.

It can be concluded that there would appear to be consensus at an international level that safety indicators other than dose and risk have a role to play in the safety case for geological disposal facilities. They could play a role for other disposal facility types, but this has not been as well developed. The prime role could be in the development and support of models for impact assessment within time frames up to 10^4 years and for providing fundamental safety arguments for the period beyond this up to 10^6 years, where comparisons could be made with the hazard potential of naturally occurring radioactive minerals such as uranium bearing ore bodies. They could also make a contribution to the arguments presenting multiple lines of reasoning, addressing aspects such as the slow movement of groundwater, the performance of natural barriers or engineered barriers made from naturally occurring materials and the time frames for transport of materials through such barriers. There may also be possibilities to support the arguments on corrosion rates of engineered containers and rates of waste form degradation.

The IAEA has been working towards developing a consensus on the use of safety indicators and reference biospheres. The BIOMASS programme has led to the development of an internationally agreed methodology for deriving reference biospheres and provides a logical and defensible development of these biospheres for use in post-closure safety assessment for near surface waste disposal facilities. Application of the methodology has been demonstrated through the development of a number of reference biospheres that can be used for purposes of illustration and as a benchmark for comparison with safety assessments performed.

REFERENCES

[1] International Atomic Energy Agency, TECDOC 767 "Safety Indicators in Different Time frames for the Safety Assessment of Underground Radioactive Waste Repositories", 1994.

[2] International Atomic Energy Agency, TECDOC 909 "Issues in Radioactive Waste Disposal", 1996.

[3] International Atomic Energy Agency, TECDOC 975 "Regulatory Decision Making in the Presence of Uncertainty in the Context of the Disposal of Long Lived Radioactive Waste", 1997.

[4] International Atomic Energy Agency, TECDOC 1077 "Critical Groups and Biospheres in the Context of Radioactive Waste Disposal", 1999.

[5] International Atomic Energy Agency, draft TECDOC "Reference Biospheres for Solid Radioactive Waste Disposal", BIOMASS Theme 1, 2001.

[6] International Atomic Energy Agency, TECDOC 1177 "Extrapolation of Short Term Observations to Time Periods Relevant to the Isolation of Long Lived Radioactive Waste", 2000.

[7] International Atomic Energy Agency, draft TECDOC "Safety Indicators, Complementary to Dose and Risk for the Assessment of Radioactive Waste Disposal", 2002.

[8] ICRP 81 Radiation Protection Recommendations on as Applied to the Disposal of Long-lived Radioactive Waste, 1999.

PART D

AGENDA

1. Technical topics

Technical Topic A: "The Different Time Scales *Versus* The Regulatory Framework and Public Acceptance".

Technical Topic B: "Barriers and System Performances within a safety case: Their Functioning and Evolution with Time".

Technical Topic C: "The Role and Limitations of Modelling in Assessing Post Closure Safety at Different Times".

Technical Topic D: "The Relative Value of Safety and Performance Indicators and Qualitative Arguments in Different Time Scales".

2. The Programme Committee

The workshop Programme Committee consists of:

Peter De Preter, Chairman	(ONDRAF/NIRAS, Belgium)
Risto Paltemaa	(STUK, Finland)
Lise Griffault	(ANDRA, France)
Didier Gay	(IRSN, France)
Klaus-Jürgen Röhlig	(GRS/Köln, Germany)
Hiroyuki Umeki	(NUMO, Japan)
Jürg Schneider	(Nagra, Switzerland)
Lucy Bailey, Anna Littleboy	(UK Nirex Ltd)
Sylvie Voinis	(OECD/NEA)
Paul A. Smith (Consultant)	(SAM Ltd, UK)

PROGRAMME

PLENARY SESSION

Experiences of Each Country in the Framework of Handling Time Scales

Chairperson: P. De Preter **(ONDRAF/NIRAS, Belgium)**
Rapporteur: P. A. Smith **(SAM Ltd, UK)**

Opening and Welcome Addresses
Chairman, P. De Preter, (ONDRAF/NIRAS, Belgium); S. Voinis, (NEA Secretariat); M. Viala, (IRSN, Host Organisation)

Some Questions on the Use of Long Time Scales for Radioactive Waste Disposal Safety Assessments
R.A. Yearsley Environment Agency, UK) and T.J. Sumerling (SAM Ltd, UK)

Handling Time Scales in Post-closure safety Assessments: A Nirex Perspective
L. Bailey and A. Littleboy (UK Nirex Ltd)

Consideration on Time Scales in the Finnish Safety Regulations for Spent Fuel Disposal
E.J. Ruokola (STUK, Finland)

Handling of Time Scales in Safety Assessments of Geological Disposal: An IRSN-GRS Standpoint on the Possible Role of Regulatory Guidance
K-J. Röhlig (GRS/Köln, Germany) and D. Gay (IRSN, France)

Handling of Time Scales in Safety Assessments: The Swiss Perspective
J.W. Schneider, P. Zuidema (Nagra, Switzerland) and P.A. Smith (SAM Ltd, UK)

Handling of Time Scales: Application of Safety Indicators
H. Umeki (NUMO, Japan)

Handling of Time Scales and Related Safety indicators
L. Griffault and E. Fillion (ANDRA, France)

Time Scales in the Long-term Safety Assessment of the Morsleben Repository, Germany
M. Ranft and J. Wollrath (BfS, Germany)

Discussion/ conclusion of the plenary session
Chairman: J. Schneider
Rapporteur: P.A. Smith

Information on the working group sessions
Chairman: P. de Preter

Parallel Working Groups Sessions on:

Topics A, B, C and D:

TOPIC A:
The Different Time Scales *Versus* The Regulatory Framework and Public Acceptance

Chairperson: J.O. Vigfusson (HSK, Switzerland)
Rapporteur: D. Gay (IRSN, France)

Long Time Scales, Low Risks: Balancing Ethics, Resources, Feasibility and Public Expectations
– N. Chapman (Nagra, Switzerland)
Safety in the First Thousand Years *– M.Jensen (SSI, Sweden)*

TOPIC B:
Barriers and System Performances within a safety case: Their Functioning and Evolution with Time

Chairperson: A. Hedin (SKB, Sweden)
Rapporteur: S. Voinis (NEA)

Fulfilment of the Long Term Safety Functions by the Different Barriers during the Main Time Frames after Reposity Closure *– P. De Preter and P. Lalieux (ONDRAF/NIRAS, Belgium)*
Treatment of Barrier Evolution: the SKB Perspective *– A.Hedin (SKB, Sweden)*

TOPIC C:
The Role and Limitations of Modelling in Assessing Post-closure Safety at Different Times

Chairperson: A. Hooper (UK Nirex Ltd)
Rapporteur: G. Ouzounian (ANDRA, France)

Phenomenology Dependent Time Scales *– G. Ouzounian (ANDRA, France)*
Long Scale Astronomical Variations in Our Solar System: Consequences for Climate *–*
K-G Karlsson and S. Edvardsson (Mid Sweden University, Sweden)

TOPIC D:
The Relative Value of Safety and Performance Indicators and Qualitative Arguments in Different Time Frames

Chairperson: J. Alonso (ENRESA, Spain)
Rapporteur: K-J. Röhlig (GRS/Köln, Germany)

Testing of Safety and Performance Indicators (SPIN) *– R. Storck and D.A. Becker (GRS/Braunschweig, Germany)*
IAEA Activities Related to Safety Indicators, Time frames and Reference Scenarios *–*
B. Batandjieva, K. Hioki and P. Metcalf (IAEA)

WORKING GROUPS SESSIONS (cont'd)

Working groups sessions (cont'd) and

Preparation of the round-up plenary session

ROUND-UP PLENARY SESSION

Chairperson: D. Sevougian (US DoE/YM)
Rapporteur: P.A. Smith (SAM Ltd, UK)

Technical Topic A   *by D. Gay*

Technical Topic B   *by S. Voinis*

Technical Topic C   *by G. Ouzounian*

Technical Topic D   *by K-J. Röhlig*

Final Discussion and Closing of the Round-up Plenary Session

Chairperson: *D. Sevougian*

Rapporteur: *P.A. Smith*

PART E

LIST OF PARTICIPANTS

BELGIUM

DE PRETER, Peter
NIRAS-ONDRAF
Avenue des Arts, 14
B-1210 Bruxelles

Tel: +32 (0)2 212 1049
Fax: +32 (0)2 218 5165
Eml: p.depreter@nirond.be

LALIEUX, Philippe
Disposal Directorate
ONDRAF/NIRAS
Avenue des Arts, 14
B-1210 Brussels

Tel: +32 (0)2 212 10 82
Fax: +32 (0)2 218 51 65
Eml: p.lalieux@nirond.be

MARIVOET, Jan
Centre d'Etude de l'Energie
Nucleaire (CEN/SCK)
Boeretang 200
B-2400 MOL

Tel: +32 (0)14 33 32 42
Fax: +32 (0)14 32 35 53
Eml: jmarivoe@sckcen.be

NYS, Vincent
Association Vinçotte Nuclear
rue Walcourt, 148
B-1070 Bruxelles

Tel: +32 (0)2 528 02 71
Fax: +32 (0)2 528 01 01
Eml: vns@avn.be

CZECH REPUBLIC

NACHMILNER, Lumir
Head, Technical Development Department
RAWRA
Dlázdená 6
110 00 Praha 1

Tel: +420 2 214 215 (ext: 19)
Fax: +420 2 214 215 44
Eml: Nachmilner@rawra.cz

FINLAND

RUOKOLA, Esko
Head of Waste Management Office
Radiation and Nuclear Safety Authority
Laippatie 4
P.O. Box 14,
FIN-00881 Helsinki

Tel: +358 9 75988305
Fax: +358 9 75988670
Eml: esko.ruokola@stuk.fi

FRANCE

FILLION, Eric
ANDRA
1-7 Rue Jean Monnet
F-92298 Chatenay Malabry

Tel: +33 (0)1 46 11 80 56
Fax: +33 (0)1 46 11 80 13
Eml: eric.fillion@andra.fr

GAY, Didier
Institut de Radioprotection et de Sûreté
Nucléaire (IRSN)
BP 17
F-92262 Fontenay-aux-Roses Cedex

Tel: +33 (0)1 58 35 91 58
Fax: +33 (0)1 58 35 77 27
Eml: didier.gay@irsn.fr

GREVOZ. Arnaud
ANDRA
Directeur Sûreté Qualité Environnement
Parc de la Croix Blanche
1-7, rue Jean Monnet
F-92298 Chatenay-Malabry Cedex

Tel: +33 1 46 11 80 35
Fax: +33 1 46 11 80 13
E-mail: arnaud.grevoz@andra.fr

GRIFFAULT, Lise
ANDRA
Parc de la Croix Blanche
1-7, rue Jean Monnet
F-92298 Chatenay-Malabry Cedex

Tel: +33 (0)1 46 11 82 96
Fax: +33 (0)1 46 11 80 63
Eml: lise.griffault@andra.fr

OUZOUNIAN, Gerald
ANDRA
Parc de la Croix Blanche
1-7, rue Jean Monnet
F-92298 Chatenay-Malabry Cedex

Tel: +33 (0)1 46 11 83 90
Fax: +33 (0)1 46 11 84 10
Eml: gerald.ouzounian@andra.fr

RAIMBAULT, Philippe
Direction Générale de la Sûreté Nucléaire
et de la Radioprotection (DGSNR)
10, Route du Panorama,
Robert Schumann – BP 83
F-92266 Fontenay-aux-Roses Cedex

Tel: +33(0)1 4319 7015
Fax: +33(0)1 4319 7166
Eml: philippe.raimbault@
        asn.minefi.gouv.fr

REGENT, Alain
Cabinet du Haut Commissaire
Commissariat à l'Energie Atomique (CEA)
31-33, rue de la Fédération
75752 Paris Cedex 15

Tel: +33 (0)1 40 56 12 19
Fax: +33 (0)1 40 56 19 75
Eml: alain.regent@cea.fr

VIALA Michele
Director for Safety of Waste Management
IRSN, BP 17
F-92262 Fontenay-aux-Roses Cedex

Tel: +33 (0)1 5835 7701
Fax: +33 (0)1 5835 79 58
E-mail: michele.viala@irsn.fr

GERMANY

KELLER Siegfried
Fed. Inst. for Geosciences and Natural Resources
Stilleweg 2
 D-30655 Hannover

Tel: +49 (511) 643 2397
Fax: +49 (511) 643 3662
E-mail: s.keller@bgr.de

LAMBERS, Ludger
Gesellschaft fuer Anlagen
und Reaktorsicherheit (GRS)
Schwertnergasse 1
D-50667 Köln

Tel: +49 (0)221 2068 797
Fax: +49 (0)221 2068 888
Eml: lal@grs.de

ROEHLIG, Klaus-Juergen
Gesellschaft für Anlagen- und
Reaktorsicherheit (GRS) mbH
Schwertnergasse 1
D-50667 Köln

Tel: +49(0)221 2068 796
Fax: +49(0)221 2068 939
Eml: rkj@grs.de

STORCK, Richard
Gesellschaft fuer Anlagen- und
Reaktorsicherheit (GRS) mbH
Theodor-Heuss-Strasse 4
D-38122 Braunscheweig

Tel: +49 (0)531 8012 205
Fax: +49 (0)531 8012 211
Eml: sto@grs.de

WOLLRATH, Jurgen
Bundesamt für Strahlenschutz (BfS)
Postfach 10 01 49
D-38201 Salzgitter

Tel: +49 (0)5341 885 642
Fax: +49 (0)5341 885 605
Eml: JWollrath@BfS.de

ITALY

SERVA, Leonello
ANPA
Via Vitaliano Brancati 48
I-00144 Roma

Tel: +39 06 5007 2539
Fax: +39 06 5007 2531
Eml: serva@anpa.it

JAPAN

UMEKI, Hiroyuki
Nuclear Waste Management Organisation
of Japan (NUMO)
Mita NN Bldg, 1-23, Shiba 4-Chome,
Minato-ku, Tokyo 108-0014

Tel: +81 (0)3 4513 1503
Fax: +81 (0)3 4513 1599
Eml: humeki@numo.or.jp

SPAIN

ALONSO, Jesus
ENRESA
Calle Emilio Vargas, 7
E-28043 Madrid

Tel: +34 91 566 8108
Fax: +34 91 566 8165
Eml: jald@enresa.es

RODRIGUEZ AREVALO, Javier
Consejo de Seguridad Nuclear
CSN
C/Justo Dorado 11
E-28040 Madrid

Tel: +34 91 346 0282
Fax: +34 91 346 0588
Eml: jra@csn.es

SWEDEN

HEDIN, Allan
Swedish Nuclear Fuel & Waste
Management Co. (SKB)
Box 5864
S-102 40 Stockholm

Tel: +46 (0)8 459 85 84
Fax: +46 (0)8 661 57 19
Eml: allan.hedin@skb.se

JENSEN, Mikael
Swedish Radiation Protection Institute (SSI)
S-171 16 Stockholm

Tel: +46 (0)8 72 97 100
Fax: +46 (0)8 72 97 162
Eml: mikael.jensen@ssi.se

KARLSSON, Karl Göran
Mitthögkolan
TFM
S-871 88 Härnösand

Tel: +46 611 86107
Fax: +46 611 86160
Eml: kg.karlsson@mh.se

WINGEFORS Stig
Swedish Nuclear Power Inspectorate (SKI)
S-106 58 Stockholm

Tel: +46 (0)8 698 84 83
Fax: +46 (0)8 661 90 86
Eml: stig.wingefors@ski.se

SWITZERLAND

CHAPMAN, Neil
Nagra
Hardstrasse, 73
CH-5430 Wettingen

Tel: +41 (0)56 437 1334
Fax: +41 (0)56 437 1207
Eml: neil.chapman@nagra.ch

SCHNEIDER, Jurg
NAGRA
Hardstrasse 73
CH-5430 Wettingen

Tel: +41 (0)56 437 1302
Fax: +41 (0)56 437 1317
Eml: schneider@nagra.ch

VIGFUSSON, Johannes O.
Section for Transport and Waste Management
HSK - Swiss Federal Nuclear Safety
Inspectorate
5232 Villigen HSK

Tel: +41(0)56 310 3974
Fax: +41 (0)56 310 3907
Eml: johannes.vigfusson@hsk.psi.ch

UNITED KINGDOM

BAILEY, Lucy
United Kingdom Nirex Ltd
Curie Avenue
Harwell, Didcot
Oxfordshire OX11 ORH

Tel: +44 (0)1235 825 357
Fax: +44 (0)1235 820 560
Eml: lucy.bailey@nirex.co.uk

HOOPER, Alan J.
Chief Scientific Advisor
United Kingdom Nirex Ldt
Curie Avenue
Harwell, Didcot
Oxfordshire OX11 0RH

Tel: +44 (0)1235 825 401
Fax: +44 (0)1235 825 289
Eml: alan.hooper@nirex.co.uk

SMITH, Paul
Safety Assessment Management Ltd (SAM Ltd)
20 Manor Place
West End
Edinburgh EH3 7DS

Tel: +44 (0)1312 266 255
Eml: paul@samltd.demon.co.uk

YEARSLEY, Roger
Environmental Policy –
Centre for Risk and Forecasting
Environment Agency
Kings Meadow House
Kings Meadow Road
Reading, RG1 8DQ

Tel: +44 (0)118 953 5258
Fax: +44 (0)118 653 5265
Eml: roger.yearsley@environment-
agency.gov.uk

UNITED STATES OF AMERICA

SEVOUGIAN David
(representing YMP)
Bechtel-SAIC, LLC.
1180 Town Center Dr.
Las Vegas, NV 89144

Tel: +1 (702) 295 6273
Fax: +1 (702) 295-0438
Eml: david_sevougian@ymp.gov

INTERNATIONAL ORGANISATIONS

BATANDJIEVA, Borislava
IAEA
Wagramerstrasse, 5
PO Box 100
1400 Vienna (Austria)

Tel: +43 1 2600 x22553
Fax: +43 1 26007
Eml: b.batandjieva@iaea.org

RIOTTE, Hans
OECD/NEA
12, Boulevard des Iles
F-92130 Issy-les-Moulineaux (France)

Tel: +33 (0)1 45 24 10 40
Fax: +33 (0)1 45 24 11 10
Eml: hans.riotte@oecd.org

VOINIS, Sylvie
OECD/NEA
12, Boulevard des Iles
F-92130 Issy-les-Moulineaux (France)

Tel: +33 (0)1 45 24 10 49
Fax: +33 (0)1 45 24 11 10
Eml: sylvie.voinis@oecd.org

ALSO AVAILABLE

NEA Publications of General Interest

2001 Annual Report (2002)

Free: paper or web.

NEA News
ISSN 1605-9581

Yearly subscription: € 37 US$ 45 GBP 26 ¥ 4 800

Radioactive Waste Management

Stepwise Decision Making in Finland for the Disposal of Spent Nuclear Fuel (2002)
ISBN 92-64-19941-1

Price: € 45 US$ 45 £ 28 ¥ 5 250

Establishing and Communicating Confidence in the Safety of Deep Geologic Disposal (2002)
ISBN 92-64-09782-1

Price: € 45 US$ 40 £ 28 ¥ 5 150

Radionuclide Retention in Geologic Media (2002)
ISBN 92-64-19695-1

Price: € 55 US$ 49 £ 34 ¥ 5 550

Using Thermodynamic Sorption Models for Guiding Radioelement Distribution Coefficient (KD) Investigations – A Status Report (2001)
ISBN 92-64-18679-4

Price: € 50 US$ 45 £ 31 ¥ 5 050

Gas Generation and Migration in Radioactive Waste Disposal – Safety-relevant Issues (2001)
ISBN 92-64-18672-7

Price: € 45 US$ 39 £ 27 ¥ 4 300

Confidence in Models of Radionuclide Transport for Site-specific Assessment (2001)
ISBN 92-64-18620-4

Price: € 96 US$ 84 £ 58 ¥ 9 100

The Decommissioning and Dismantling of nuclear Facilities (2002)
ISBN 92-64-18488-0

Free: paper or web.

An International Peer Review of the Yucca Mountain Project TSPA-SR (2002)
ISBN 92-64-18477-5

Free: paper or web.

GEOTRAP: Radionuclide Migration in Geologic Heterogeneous Media (2002)
ISBN 92-64-18479-1

Free: paper or web.

The Role of Underground Laboratories in Nuclear Waste Disposal Programmes (2001)
ISBN 92-64-18472-4

Free: paper or web.

Nuclear Waste Bulletin – Update on Waste Management Policies and Programmes, No 14, 2000 Edition (2001)
ISBN 92-64-18461-9

Free: paper or web.

Order form on reverse side.

ORDER FORM

OECD Nuclear Energy Agency, 12 boulevard des Iles, F-92130 Issy-les-Moulineaux, France
Tel. 33 (0)1 45 24 10 15, Fax 33 (0)1 45 24 11 10, E-mail: nea@nea.fr, Internet: www.nea.fr

Qty	Title	ISBN	Price	Amount
			Total	

Charge my credit card ❑ VISA ❑ Mastercard ❑ Eurocard ❑ American Express

Card No.	Expiration date	Signature
Name		
Address	Country	
Telephone	Fax	
E-mail		

OECD PUBLICATIONS, 2, rue André-Pascal, 75775 PARIS CEDEX 16
PRINTED IN FRANCE
(66 2002 18 1 P) ISBN 92-64-09911-5 – No. 52787 2002